# Video Codec Design

To
Freya and Hugh

# Video Codec Design

## Developing Image and Video Compression Systems

**Iain E. G. Richardson**

*The Robert Gordon University, Aberdeen, UK*

JOHN WILEY & SONS, LTD

***Other Wiley Editorial Offices***

John Wiley & Sons Inc., 111 River Street, Hoboken, NJ 07030, USA

Jossey-Bass, 989 Market Street, San Francisco, CA 94103-1741, USA

Wiley-VCH Verlag GmbH, Boschstr. 12, D-69469 Weinheim, Germany

John Wiley & Sons Australia Ltd, 33 Park Road, Milton, Queensland 4064, Australia

John Wiley & Sons (Asia) Pte Ltd, 2 Clementi Loop #02-01, Jin Xing Distripark, Singapore 129809

John Wiley & Sons Canada Ltd, 22 Worcester Road, Etobicoke, Ontario, Canada M9W 1L1

***British Library Cataloguing in Publication Data***

A catalogue record for this book is available from the British Library

ISBN 0 471 48553 5

Typeset in 10/12pt Times by Thomson Press (India) Ltd, New Delhi
Printed and bound in Great Britain by Antony Rowe Ltd, Chippenham, Wiltshire
This book is printed on acid-free paper responsibly manufactured from sustainable forestry
in which at least two trees are planted for each one used for paper production.

# Contents

# 1

# Introduction

## 1.1 IMAGE AND VIDEO COMPRESSION

The subject of this book is the compression ('coding') of digital images and video. Within the last 5–10 years, image and video coding have gone from being relatively esoteric research subjects with few 'real' applications to become key technologies for a wide range of mass-market applications, from personal computers to television.

Like many other recent technological developments, the emergence of video and image coding in the mass market is due to convergence of a number of areas. Cheap and powerful processors, fast network access, the ubiquitous Internet and a large-scale research and standardisation effort have all contributed to the development of image and video coding technologies. Coding has enabled a host of new 'multimedia' applications including digital television, digital versatile disk (DVD) movies, streaming Internet video, home digital photography and video conferencing.

Compression coding bridges a crucial gap in each of these applications: the gap between the user's demands (high-quality still and moving images, delivered quickly at a reasonable cost) and the limited capabilities of transmission networks and storage devices. For example, a 'television-quality' digital video signal requires 216 Mbits of storage or transmission capacity for one second of video. Transmission of this type of signal in real time is beyond the capabilities of most present-day communications networks. A 2-hour movie (uncompressed) requires over 194 Gbytes of storage, equivalent to 42 DVDs or 304 CD-ROMs. In order for digital video to become a plausible alternative to its analogue predecessors (analogue television or VHS videotape), it has been necessary to develop methods of reducing or compressing this prohibitively high bit-rate signal.

The drive to solve this problem has taken several decades and massive efforts in research, development and standardisation (and work continues to improve existing methods and develop new coding paradigms). However, efficient compression methods are now a firmly established component of the new digital media technologies such as digital television and DVD-video. A welcome side effect of these developments is that video and image compression has enabled many novel visual communication applications that would not have previously been possible. Some areas have taken off more quickly than others (for example, the long-predicted boom in video conferencing has yet to appear), but there is no doubt that visual compression is here to stay. Every new PC has a number of designed-in features specifically to support and accelerate video compression algorithms. Most developed nations have a timetable for stopping the transmission of analogue television, after which all television receivers will need compression technology to decode and display TV images. VHS videotapes are finally being replaced by DVDs which can be played back on

DVD players or on PCs. The heart of all of these applications is the video compressor and decompressor; or enCOder/DECoder; or *video CODEC*.


## 1.2   VIDEO CODEC DESIGN

Video CODEC technology has in the past been something of a 'black art' known only to a small community of academics and technical experts, partly because of the lack of approachable, practical literature on the subject. One view of image and video coding is as a mathematical process. The video coding field poses a number of interesting mathematical problems and this means that much of the literature on the subject is, of necessity, highly mathematical. Such a treatment is important for developing the fundamental concepts of compression but can be bewildering for an engineer or developer who wants to put compression into practice. The increasing prevalence of digital video applications has led to the publication of more approachable texts on the subject: unfortunately, some of these offer at best a superficial treatment of the issues, which can be equally unhelpful.

This book aims to fill a gap in the market between theoretical and over-simplified texts on video coding. It is written primarily from a design and implementation perspective. Much work has been done over the last two decades in developing a portfolio of practical techniques and approaches to video compression coding as well as a large body of theoretical research. A grasp of these design techniques, trade-offs and performance issues is important to anyone who needs to design, specify or interface to video CODECs. This book emphasises these practical considerations rather than rigorous mathematical theory and concentrates on the current generation of video coding systems, embodied by the MPEG-2, MPEG-4 and H.263 standards. By presenting the practicalities of video CODEC design in an approachable way it is hoped that this book will help to demystify this important technology.


## 1.3   STRUCTURE OF THIS BOOK

The book is organised in three main sections (Figure 1.1). We deal first with the fundamental concepts of digital video, image and video compression and the main international standards for video coding (Chapters 2–5). The second section (Chapters 6–9) covers the key components of video CODECs in some detail. Finally, Chapters 10–14 discuss system design issues and present some design case studies.

Chapter 2, 'Digital Video', explains the concepts of video capture, representation and display; discusses the way in which we perceive visual information; compares methods for measuring and evaluate visual 'quality'; and lists some applications of digital video.

Chapter 3, 'Image and Video Compression Fundamentals', examines the requirements for video and image compression and describes the components of a 'generic' image CODEC and video CODEC. (Note: this chapter deliberately avoids discussing technical or standard-specific details of image and video compression.)

Chapter 4, 'JPEG and MPEG', describes the operation of the international standards bodies and introduces the ISO image and video compression standards: JPEG, Motion JPEG and JPEG-2000 for images and MPEG-1, MPEG-2 and MPEG-4 for moving video.

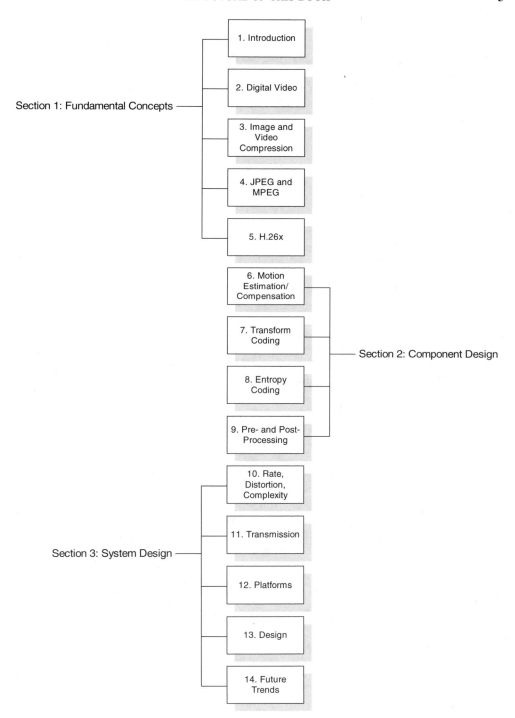

**Figure 1.1** Structure of the book

Chapter 5, 'H.261, H.263 and H.26L', explains the concepts of the ITU-T video coding standards H.261 and H.263 and the emerging H.26L. The chapter ends with a comparison of the performance of the main image and video coding standards.

Chapter 6, 'Motion Estimation and Compensation', deals with the 'front end' of a video CODEC. The requirements and goals of motion-compensated prediction are explained and the chapter discusses a number of practical approaches to motion estimation in software or hardware designs.

Chapter 7, 'Transform Coding', concentrates mainly on the popular discrete cosine transform. The theory behind the DCT is introduced and practical algorithms for calculating the forward and inverse DCT are described. The discrete wavelet transform (an increasingly popular alternative to the DCT) and the process of quantisation (closely linked to transform coding) are discussed.

Chapter 8, 'Entropy Coding', explains the statistical compression process that forms the final step in a video encoder; shows how Huffman code tables are designed and used; introduces arithmetic coding; and describes practical entropy encoder and decoder designs.

Chapter 9, 'Pre- and Post-processing', addresses the important issue of input and output processing; shows how pre-filtering can improve compression performance; and examines a number of post-filtering techniques, from simple de-blocking filters to computationally complex, high-performance algorithms.

Chapter 10, 'Rate, Distortion and Complexity', discusses the relationships between compressed bit rate, visual distortion and computational complexity in a 'lossy' video CODEC; describes rate control algorithms for different transmission environments; and introduces the emerging techniques of variable-complexity coding that allow the designer to trade computational complexity against visual quality.

Chapter 11, 'Transmission of Coded Video', addresses the influence of the transmission environment on video CODEC design; discusses the quality of service required by a video CODEC and provided by typical transport scenarios; and examines ways in which quality of service can be 'matched' between the CODEC and the network to maximise visual quality.

Chapter 12, 'Platforms', describes a number of alternative platforms for implementing practical video CODECs, ranging from general-purpose PC processors to custom-designed hardware platforms.

Chapter 13, 'Video CODEC Design', brings together a number of the themes discussed in previous chapters and discusses how they influence the design of video CODECs; examines the interfaces between a video CODEC and other system components; and presents two design studies, a software CODEC and a hardware CODEC.

Chapter 14, 'Future Developments', summarises some of the recent work in research and development that will influence the next generation of video CODECs.

Each chapter includes references to papers and websites that are relevant to the topic. The bibliography lists a number of books that may be useful for further reading and a companion web site to the book may be found at:

http://www.vcodex.com/videocodecdesign/

# 2
# Digital Video

## 2.1 INTRODUCTION

Digital video is now an integral part of many aspects of business, education and entertainment, from digital TV to web-based video news. Before examining methods for compressing and transporting digital video, it is necessary to establish the concepts and terminology relating to video in the digital domain. *Digital video* is visual information represented in a discrete form, suitable for digital electronic storage and/or transmission. In this chapter we describe and define the concept of digital video: essentially a sampled two-dimensional (2-D) version of a continuous three-dimensional (3-D) scene. Dealing with colour video requires us to choose a colour space (a system for representing colour) and we discuss two widely used colour spaces, RGB and YCrCb. The goal of a video coding system is to support video communications with an 'acceptable' visual quality: this depends on the viewer's perception of visual information, which in turn is governed by the behaviour of the human visual system. Measuring and quantifying visual quality is a difficult problem and we describe some alternative approaches, from time-consuming subjective tests to automatic objective tests (with varying degrees of accuracy).

## 2.2 CONCEPTS, CAPTURE AND DISPLAY

### 2.2.1 The Video Image

A video image is a projection of a 3-D scene onto a 2-D plane (Figure 2.1). A 3-D scene consisting of a number of objects each with depth, texture and illumination is projected onto a plane to form a 2-D representation of the scene. The 2-D representation contains varying texture and illumination but no depth information. A *still image* is a 'snapshot' of the 2-D representation at a particular instant in time whereas a *video sequence* represents the scene over a period of time.

### 2.2.2 Digital Video

A 'real' visual scene is continuous both spatially and temporally. In order to represent and process a visual scene digitally it is necessary to sample the real scene spatially (typically on a rectangular grid in the video image plane) and temporally (typically as a series of 'still'

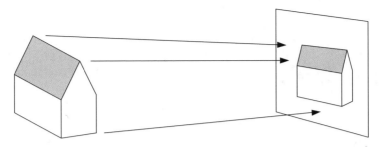

**Figure 2.1**   Projection of 3-D scene onto a video image

# Moving scene

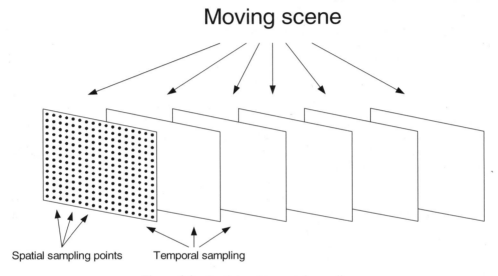

Spatial sampling points        Temporal sampling

**Figure 2.2**   Spatial and temporal sampling

images or *frames* sampled at regular intervals in time) as shown in Figure 2.2. Digital video is the representation of a spatio-temporally sampled video scene in digital form. Each spatio-temporal sample (described as a picture element or *pixel*) is represented digitally as one or more numbers that describe the brightness (*luminance*) and colour of the sample.

A *digital video system* is shown in Figure 2.3. At the input to the system, a 'real' visual scene is captured, typically with a camera and converted to a sampled digital representation.

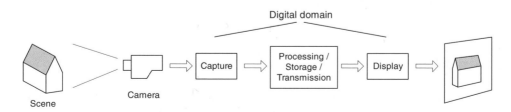

**Figure 2.3**   Digital video system: capture, processing and display

This digital video signal may then be handled in the digital domain in a number of ways, including processing, storage and transmission. At the output of the system, the digital video signal is displayed to a viewer by reproducing the 2-D video image (or video sequence) on a 2-D display.

### 2.2.3 Video Capture

Video is captured using a camera or a system of cameras. Most current digital video systems use 2-D video, captured with a single camera. The camera focuses a 2-D projection of the video scene onto a sensor, such as an array of charge coupled devices (CCD array). In the case of colour image capture, each colour component (see Section 2.3) is filtered and projected onto a separate CCD array.

Figure 2.4 shows a two-camera system that captures two 2-D projections of the scene, taken from different viewing angles. This provides a stereoscopic representation of the scene: the two images, when viewed in the left and right eye of the viewer, give an appearance of 'depth' to the scene. There is an increasing interest in the use of 3-D digital video, where the video signal is represented and processed in three dimensions. This requires the capture system to provide depth information as well as brightness and colour, and this may be obtained in a number of ways. Stereoscopic images can be processed to extract approximate depth information and form a 3-D representation of the scene: other methods of obtaining depth information include processing of multiple images from a single camera (where either the camera or the objects in the scene are moving) and the use of laser 'striping' to obtain depth maps. In this book we will concentrate on 2-D video systems.

Generating a digital representation of a video scene can be considered in two stages: acquisition (converting a projection of the scene into an electrical signal, for example via a CCD array) and digitisation (sampling the projection spatially and temporally and converting each sample to a number or set of numbers). Digitisation may be carried out using a separate device or board (e.g. a video capture card in a PC): increasingly, the digitisation process is becoming integrated with cameras so that the output of a camera is a signal in sampled digital form.

### 2.2.4 Sampling

A digital image may be generated by sampling an analogue video signal (i.e. a varying electrical signal that represents a video image) at regular intervals. The result is a sampled

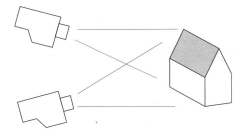

**Figure 2.4** Stereoscopic camera system

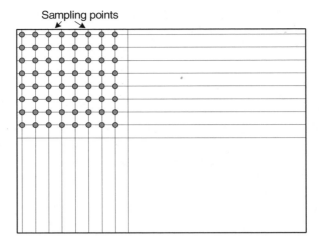

**Figure 2.5**  Spatial sampling (square grid)

version of the image: the sampled image is only defined at a series of regularly spaced sampling points. The most common format for a sampled image is a rectangle (often with width larger than height) with the sampling points positioned on a square grid (Figure 2.5). The visual quality of the image is influenced by the number of sampling points. More sampling points (a higher sampling resolution) give a 'finer' representation of the image: however, more sampling points require higher storage capacity. Table 2.1 lists some commonly used image resolutions and gives an approximately equivalent analogue video quality: VHS video, broadcast TV and high-definition TV.

A moving video image is formed by sampling the video signal temporally, taking a rectangular 'snapshot' of the signal at periodic time intervals. Playing back the series of frames produces the illusion of motion. A higher temporal sampling rate (*frame rate*) gives a 'smoother' appearance to motion in the video scene but requires more samples to be captured and stored (see Table 2.2). Frame rates below 10 frames per second are sometimes

**Table 2.1**  Typical video image resolutions

| Image resolution | Number of sampling points | Analogue video 'equivalent' |
| --- | --- | --- |
| $352 \times 288$ | 101 376 | VHS video |
| $704 \times 576$ | 405 504 | Broadcast television |
| $1440 \times 1152$ | 1 313 280 | High-definition television |

**Table 2.2**  Video frame rates

| Video frame rate | Appearance |
| --- | --- |
| Below 10 frames per second | 'Jerky', unnatural appearance to movement |
| 10–20 frames per second | Slow movements appear OK; rapid movement is clearly 'jerky' |
| 20–30 frames per second | Movement is reasonably smooth |
| 50–60 frames per second | Movement is very smooth |

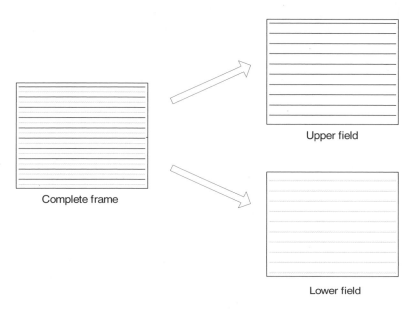

Upper field

Complete frame

Lower field

**Figure 2.6** Interlaced fields

used for very low bit-rate video communications (because the amount of data is relatively small): however, motion is clearly jerky and unnatural at this rate. Between 10 and 20 frames per second is more typical for low bit-rate video communications; 25 or 30 frames per second is standard for television pictures (together with the use of interlacing, see below); 50 or 60 frames per second is appropriate for high-quality video (at the expense of a very high data rate).

The visual appearance of a temporally sampled video sequence can be improved by using *interlaced video*, commonly used for broadcast-quality television signals. For example, the European PAL video standard operates at a temporal frame rate of 25 Hz (i.e. 25 complete frames of video per second). However, in order to improve the visual appearance without increasing the data rate, the video sequence is composed of *fields* at a rate of 50 Hz (50 fields per second). Each field contains half of the lines that make up a complete frame (Figure 2.6): the odd- and even-numbered lines from the frame on the left are placed in two separate fields, each containing half the information of a complete frame. These fields are captured and displayed at 1/50th of a second intervals and the result is an update rate of 50 Hz, with the data rate of a signal at 25 Hz. Video that is captured and displayed in this way is known as interlaced video and generally has a more pleasing visual appearance than video transmitted as complete frames (*non-interlaced* or *progressive video*). Interlaced video can, however, produce unpleasant visual artefacts when displaying certain textures or types of motion.

## 2.2.5 Display

Displaying a 2-D video signal involves recreating each frame of video on a 2-D display device. The most common type of display is the cathode ray tube (CRT) in which the image

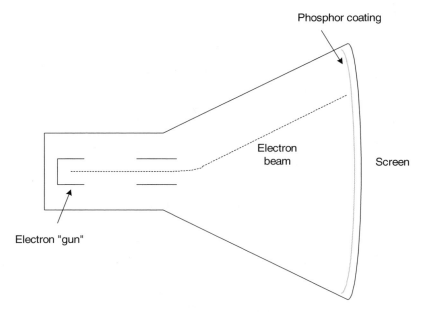

**Figure 2.7**   CRT display

is formed by scanning a modulated beam of electrons across a phosphorescent screen (Figure 2.7). CRT technology is mature and reasonably cheap to produce. However, a CRT suffers from the requirement to provide a sufficiently long path for the electron beam (making the device bulky) and the weight of the vacuum tube. Liquid crystal displays (LCDs) are becoming a popular alternative to the CRT for computer applications but are not as bright; other alternatives such as flat-panel plasma displays are beginning to emerge but are not yet available at competitive prices.

## 2.3   COLOUR SPACES

A monochrome ('grey scale') video image may be represented using just one number per spatio-temporal sample. This number indicates the brightness or luminance of each sample position: conventionally, a larger number indicates a brighter sample. If a sample is represented using $n$ bits, then a value of 0 may represent black and a value of $(2^n - 1)$ may represent white, with other values in between describing shades of grey. Luminance is commonly represented with 8 bits per sample for 'general-purpose' video applications. Higher luminance 'depths' (e.g. 12 bits or more per sample) are sometimes used for specialist applications (such as digitising of X-ray slides).

Representing colour requires multiple numbers per sample. There are several alternative systems for representing colour, each of which is known as a *colour space*. We will concentrate here on two of the most common colour spaces for digital image and video representation: RGB (red/green/blue) and YCrCb (luminance/red chrominance/blue chrominance).

## 2.3.1 RGB

In the red/green/blue colour space, each pixel is represented by three numbers indicating the relative proportions of red, green and blue. These are the three additive primary colours of light: any colour may be reproduced by combining varying proportions of red, green and blue light. Because the three components have roughly equal importance to the final colour, RGB systems usually represent each component with the same precision (and hence the same number of bits). Using 8 bits per component is quite common: $3 \times 8 = 24$ bits are required to represent each pixel. Figure 2.8 shows an image (originally colour, but displayed here in monochrome!) and the brightness 'maps' of each of its three colour components. The girl's cap is a bright pink colour: this appears bright in the red component and slightly less bright in the blue component.

(a)

(b)

**Figure 2.8** (a) Image, (b) R, (c) G, (d) B components

(c)

(d)

**Figure 2.8**   (*Continued*)

## 2.3.2   YCrCb

RGB is not necessarily the most efficient representation of colour. The human visual system (HVS, see Section 2.4) is less sensitive to colour than to luminance (brightness): however, the RGB colour space does not provide an easy way to take advantage of this since the three colours are equally important and the luminance is present in all three colour components. It is possible to represent a colour image more efficiently by separating the luminance from the colour information.

A popular colour space of this type is Y: Cr: Cb. Y is the luminance component, i.e. a monochrome version of the colour image. Y is a weighted average of R, G and B:

$$Y = k_r R + k_g G + k_b B$$

where $k$ are weighting factors. The colour information can be represented as *colour difference* or *chrominance* components, where each chrominance component is the difference between R, G or B and the luminance Y:

$$Cr = R - Y$$
$$Cb = B - Y$$
$$Cg = G - Y$$

The complete description is given by Y (the luminance component) and three colour differences Cr, Cb and Cg that represent the 'variation' between the colour intensity and the 'background' luminance of the image.

So far, this representation has little obvious merit: we now have four components rather than three. However, it turns out that the value of $Cr + Cb + Cg$ is a constant. This means that only two of the three chrominance components need to be transmitted: the third component can always be found from the other two. In the Y : Cr : Cb space, only the luminance (Y) and red and blue chrominance (Cr, Cb) are transmitted. Figure 2.9 shows the effect of this operation on the colour image. The two chrominance components only have significant values where there is a significant 'presence' or 'absence' of the appropriate colour (for example, the pink hat appears as an area of relative brightness in the red chrominance).

The equations for converting an RGB image into the Y : Cr : Cb colour space and vice versa are given in Equations 2.1 and 2.2. Note that G can be extracted from the Y : Cr : Cb representation by subtracting Cr and Cb from Y.

$$Y = 0.299\,R + 0.587\,G + 0.114\,B$$
$$Cb = 0.564\,(B - Y) \tag{2.1}$$
$$Cr = 0.713\,(R - Y)$$

$$R = Y + 1.402\,Cr$$
$$G = Y - 0.344\,Cb - 0.714\,Cr \tag{2.2}$$
$$B = Y + 1.772\,Cb$$

The key advantage of Y : Cr : Cb over RGB is that the Cr and Cb components may be represented with a *lower resolution* than Y because the HVS is less sensitive to colour than luminance. This reduces the amount of data required to represent the chrominance components without having an obvious effect on visual quality: to the casual observer, there is no apparent difference between an RGB image and a Y : Cr : Cb image with reduced chrominance resolution.

Figure 2.10 shows three popular 'patterns' for sub-sampling Cr and Cb. 4 : 4 : 4 means that the three components (Y : Cr : Cb) have the same resolution and hence a sample of each component exists at every pixel position. (The numbers indicate the relative sampling rate of each component in the *horizontal* direction, i.e. for every 4 luminance samples there are 4 Cr and 4 Cb samples.) 4 : 4 : 4 sampling preserves the full fidelity of the chrominance components. In 4 : 2 : 2 sampling, the chrominance components have the same vertical resolution but half the horizontal resolution (the numbers indicate that for every 4 luminance

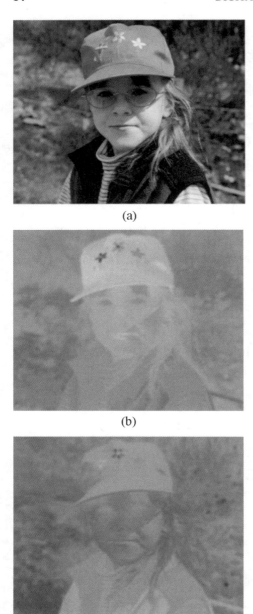

(a)

(b)

(c)          **Figure 2.9**    (a) Luminance, (b) Cr, (c) Cb components

samples in the horizontal direction there are 2 Cr and 2 Cb samples) and the locations of the samples are shown in the figure. 4 : 2 : 2 video is used for high-quality colour reproduction.

4 : 2 : 0 means that Cr and Cb each have half the horizontal and vertical resolution of Y, as shown. The term '4 : 2 : 0' is rather confusing: the numbers do not actually have a sensible interpretation and appear to have been chosen historically as a 'code' to identify this

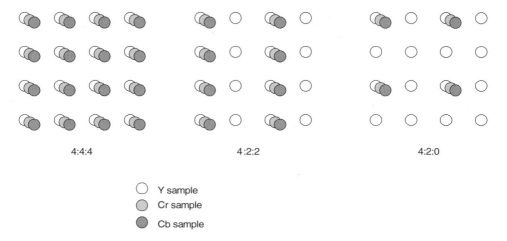

4:4:4          4:2:2          4:2:0

○ Y sample
◔ Cr sample
● Cb sample

**Figure 2.10**  Chrominance subsampling patterns

particular sampling pattern. $4:2:0$ sampling is popular in 'mass market' digital video applications such as video conferencing, digital television and DVD storage. Because each colour difference component contains a quarter of the samples of the Y component, $4:2:0$ video requires exactly half as many samples as $4:4:4$ (or $R:G:B$) video.

*Example*

Image resolution: $720 \times 576$ pixels
Y resolution: $720 \times 576$ samples, each represented with 8 bits

$4:4:4$ Cr, Cb resolution: $720 \times 576$ samples, each 8 bits
Total number of bits: $720 \times 576 \times 8 \times 3 = 9\,953\,280$ bits

$4:2:0$ Cr, Cb resolution: $360 \times 288$ samples, each 8 bits
Total number of bits: $(720 \times 576 \times 8) + (360 \times 288 \times 8 \times 2) = 4\,976\,640$ bits

The $4:2:0$ version requires half as many bits as the $4:4:4$ version

To further confuse things, $4:2:0$ sampling is sometimes described as '12 bits per pixel'. The reason for this can be illustrated by examining a group of 4 pixels (Figure 2.11). The left-hand diagram shows $4:4:4$ sampling: a total of 12 samples are required, 4 each of Y, Cr and Cb, requiring a total of $12 \times 8 = 96$ bits, i.e. an average of $96/4 = 24$ bits per pixel. The right-hand diagram shows $4:2:0$ sampling: 6 samples are required, 4 Y and one each of Cr, Cb, requiring a total of $6 \times 8 = 48$ bits, i.e. an average of $48/4 = 12$ bits per pixel.

**Figure 2.11**  4 pixels: 24 and 12 bpp

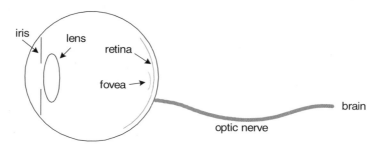

**Figure 2.12**  HVS components

## 2.4  THE HUMAN VISUAL SYSTEM

A critical design goal for a digital video system is that the visual images produced by the system should be 'pleasing' to the viewer. In order to achieve this goal it is necessary to take into account the response of the *human visual system* (HVS). The HVS is the 'system' by which a human observer views, interprets and responds to visual stimuli. The main components of the HVS are shown in Figure 2.12:

- *Eye*: The image is focused by the *lens* onto the photodetecting area of the eye, the *retina*. Focusing and object tracking are achieved by the eye *muscles* and the *iris* controls the aperture of the lens and hence the amount of light entering the eye.

- *Retina*: The retina consists of an array of *cones* (photoreceptors sensitive to colour at high light levels) and *rods* (photoreceptors sensitive to luminance at low light levels). The more sensitive cones are concentrated in a central region (the *fovea*) which means that high-resolution colour vision is only achieved over a small area at the centre of the field of view.

- *Optic nerve*: This carries electrical signals from the retina to the brain.

- *Brain*: The human brain processes and interprets visual information, based partly on the received information (the image detected by the retina) and partly on prior learned responses (such as known object shapes).

The operation of the HVS is a large and complex area of study. Some of the important features of the HVS that have implications for digital video system design are listed in Table 2.3.

## 2.5  VIDEO QUALITY

In order to specify, evaluate and compare video communication systems it is necessary to determine the quality of the video images displayed to the viewer. Measuring visual quality is a difficult and often imprecise art because there are so many factors that can influence the results. Visual quality is inherently *subjective* and is therefore influenced by many subjective factors that can make it difficult to obtain a completely accurate measure of quality.

**Table 2.3**  Features of the HVS

| Feature | Implication for digital video systems |
|---|---|
| The HVS is more sensitive to luminance detail than to colour detail | Colour (or chrominance) resolution may be reduced without significantly affecting image quality |
| The HVS is more sensitive to high contrast (i.e. large differences in luminance) than low contrast | Large changes in luminance (e.g. edges in an image) are particularly important to the appearance of the image |
| The HVS is more sensitive to low spatial frequencies (i.e. changes in luminance that occur over a large area) than high spatial frequencies (rapid changes that occur in a small area) | It may be possible to compress images by discarding some of the less important higher frequencies (however, edge information should be preserved) |
| The HVS is more sensitive to image features that persist for a long duration | It is important to minimise temporally persistent disturbances or artefacts in an image |
| The illusion of 'smooth' motion can be achieved by presenting a series of images at a rate of 20–30 Hz or more | Video systems should aim for frame repetition rates of 20 Hz or more for 'natural' moving video |
| HVS responses vary from individual to individual | Multiple observers should be used to assess the quality of a video system |

Measuring visual quality using *objective* criteria gives accurate, repeatable results, but as yet there are no objective measurement systems that will completely reproduce the subjective experience of a human observer watching a video display.

## 2.5.1   Subjective Quality Measurement

Several test procedures for subjective quality evaluation are defined in ITU-R Recommendation BT.500-10.[1] One of the most popular of these quality measures is the double stimulus continuous quality scale (DSCQS) method. An assessor is presented with a pair of images or short video sequences A and B, one after the other, and is asked to give A and B a 'score' by marking on a continuous line with five intervals. Figure 2.13 shows an example of the rating form on which the assessor grades each sequence.

In a typical test session, the assessor is shown a series of sequence pairs and is asked to grade each pair. Within each pair of sequences, one is an unimpaired 'reference' sequence and the other is the same sequence, modified by a system or process under test. A typical example from the evaluation of video coding systems is shown in Figure 2.14: the original sequence is compared with the same sequence, encoded and decoded using a video CODEC.

The order of the two sequences, original and 'impaired', is randomised during the test session so that the assessor does not know which is the original and which is the impaired sequence. This helps prevent the assessor from prejudging the impaired sequence compared with the reference sequence. At the end of the session, the scores are converted to a normalised range and the result is a score (sometimes described as a 'mean opinion score') that indicates the *relative* quality of the impaired and reference sequences.

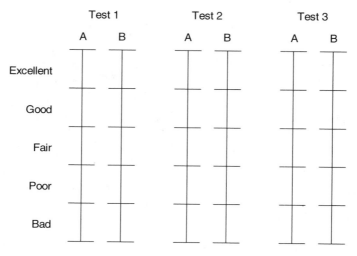

**Figure 2.13**   DSCQS rating form

The DSCQS test is generally accepted as a realistic measure of subjective visual quality. However, it suffers from practical problems. The results can vary significantly, depending on the assessor and also on the video sequence under test. This variation can be compensated for by repeating the test with several sequences and several assessors. An 'expert' assessor (e.g. one who is familiar with the nature of video compression distortions or 'artefacts') may give a biased score and it is preferable to use 'non-expert' assessors. In practice this means that a large pool of assessors is required because a non-expert assessor will quickly learn to recognise characteristic artefacts in the video sequences. These factors make it expensive and time-consuming to carry out the DSCQS tests thoroughly.

A second problem is that this test is only really suitable for short sequences of video. It has been shown[2] that the 'recency effect' means that the viewer's opinion is heavily biased towards the last few seconds of a video sequence: the quality of this last section will strongly influence the viewer's rating for the whole of a longer sequence. Subjective tests are also influenced by the viewing conditions: a test carried out in a comfortable, relaxed environment will earn a higher rating than the same test carried out in a less comfortable setting.

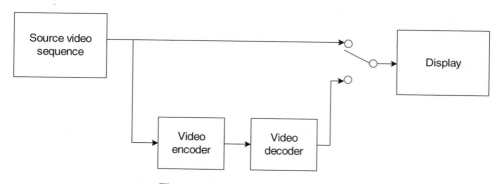

**Figure 2.14**   DSCQS testing system

## 2.5.2   Objective Quality Measurement

Because of the problems of subjective measurement, developers of digital video systems rely heavily on objective measures of visual quality. Objective measures have not yet replaced subjective testing: however, they are considerably easier to apply and are particularly useful during development and for comparison purposes.

Probably the most widely used objective measure is peak signal to noise ratio (PSNR), calculated using Equation 2.3. PSNR is measured on a logarithmic scale and is based on the mean squared error (MSE) between an original and an impaired image or video frame, relative to $(2^n - 1)^2$ (the square of the highest possible signal value in the image).

$$\text{PSNR}_{dB} = 10 \log_{10} \frac{(2^n - 1)^2}{\text{MSE}} \qquad (2.3)$$

PSNR can be calculated very easily and is therefore a very popular quality measure. It is widely used as a method of comparing the 'quality' of compressed and decompressed video images. Figure 2.15 shows some examples: the first image (a) is the original and (b), (c) and (d) are compressed and decompressed versions of the original image. The progressively poorer image quality is reflected by a corresponding drop in PSNR.

The PSNR measure suffers from a number of limitations, however. PSNR requires an 'unimpaired' original image for comparison: this may not be available in every case and it may not be easy to verify that an 'original' image has perfect fidelity. A more important limitation is that PSNR does not correlate well with subjective video quality measures such as ITU-R 500. For a given image or image sequence, high PSNR indicates relatively high quality and low PSNR indicates relatively low quality. However, a particular value of PSNR does not necessarily equate to an 'absolute' subjective quality. For example, Figure 2.16 shows two impaired versions of the original image from Figure 2.15. Image (a) (with a blurred background) has a PSNR of 32.7 dB, whereas image (b) (with a blurred foreground) has a higher PSNR of 37.5 dB. Most viewers would rate image (b) as significantly poorer than image (a): however, the PSNR measure simply counts the mean squared pixel errors and by this method image (b) is ranked as 'better' than image (a). This example shows that PSNR ratings do not necessarily correlate with 'true' subjective quality.

Because of these problems, there has been a lot of work in recent years to try to develop a more sophisticated objective test that closely approaches subjective test results. Many different approaches have been proposed,[3-5] but none of these has emerged as clear alternatives to subjective tests. With improvements in objective quality measurement, however, some interesting applications become possible, such as proposals for 'constant-quality' video coding[6] (see Chapter 10, 'Rate Control').

ITU-R BT.500-10 (and more recently, P.910) describe standard methods for subjective quality evaluation: however, as yet there is no standardised, accurate system for objective ('automatic') quality measurement that is suitable for digitally coded video. In recognition of this, the ITU-T Video Quality Experts Group (VQEG) are developing a standard for objective video quality evaluation[7]. The first step in this process was to test and compare potential models for objective evaluation. In March 2000, VQEG reported on the first round of tests in which 10 competing systems were tested under identical conditions.

(a)

(b)

**Figure 2.15**    PSNR examples: (a) original; (b) 33.2 dB; (c) 31.8 dB; (d) 26.5 dB

(c)

(d)

**Figure 2.15**   (*Continued*)

(a)

(b)

**Figure 2.16**    (a) Impairment 1 (32.7 dB); (b) impairment 2 (37.5 dB)

**Table 2.4**   ITU-R BT.601-5 parameters

|                                   | 30 Hz frame rate | 25 Hz frame rate |
|-----------------------------------|:---------------:|:---------------:|
| Fields per second                 | 60              | 50              |
| Lines per complete frame          | 525             | 625             |
| Luminance samples per line        | 858             | 864             |
| Chrominance samples per line      | 429             | 432             |
| Bits per sample                   | 8               | 8               |
| Total bit rate                    | 216 Mbps        | 216 Mbps        |
| Active lines per frame            | 480             | 576             |
| Active samples per line (Y)       | 720             | 720             |
| Active samples per line (Cr, Cb)  | 360             | 360             |

Unfortunately, none of the 10 proposals was considered suitable for standardisation. The problem of accurate objective quality measurement is therefore likely to remain for some time to come.

The PSNR measure is widely used as an approximate objective measure for visual quality and so we will use this measure for quality comparison in this book. However, it is worth remembering the limitations of PSNR when comparing different systems and techniques.

## 2.6   STANDARDS FOR REPRESENTING DIGITAL VIDEO

A widely used format for digitally coding video signals for television production is ITU-R Recommendation BT.601-5[8] (the term 'coding' in this context means conversion to digital format and does not imply compression). The luminance component of the video signal is sampled at 13.5 MHz and the chrominance at 6.75 MHz to produce a $4:2:2$ Y:Cr:Cb component signal. The parameters of the sampled digital signal depend on the video frame rate (either 30 or 25 Hz) and are shown in Table 2.4. It can be seen that the higher 30 Hz frame rate is compensated for by a lower spatial resolution so that the total bit rate is the same in each case (216 Mbps). The actual area shown on the display, the *active area*, is smaller than the total because it excludes horizontal and vertical blanking intervals that exist 'outside' the edges of the frame. Each sample has a possible range of 0–255: however, levels of 0 and 255 are reserved for synchronisation. The active luminance signal is restricted to a range of 16 (black) to 235 (white).

For video coding applications, video is often converted to one of a number of 'intermediate formats' prior to compression and transmission. A set of popular frame resolutions is based around the common intermediate format, CIF, in which each frame has a

**Table 2.5**   Intermediate formats

| Format             | Luminance resolution (horiz. × vert.) |
|--------------------|:-------------------------------------:|
| Sub-QCIF           | $128 \times 96$                       |
| Quarter CIF (QCIF) | $176 \times 144$                      |
| CIF                | $352 \times 288$                      |
| 4CIF               | $704 \times 576$                      |

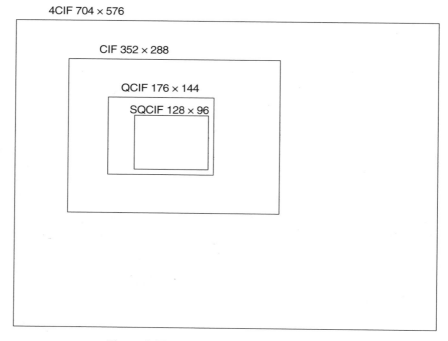

**Figure 2.17**   Intermediate formats (illustration)

resolution of 352 × 288 pixels. The resolutions of these formats are listed in Table 2.5 and their relative dimensions are illustrated in Figure 2.17.

## 2.7   APPLICATIONS

The last decade has seen a rapid increase in applications for digital video technology and new, innovative applications continue to emerge. A small selection is listed here:

- *Home video*: Video camera recorders for professional and home use are increasingly moving away from analogue tape to digital media (including digital storage on tape and on solid-state media). Affordable DVD video recorders will soon be available for the home.

- *Video storage*: A variety of digital formats are now used for storing video on disk, tape and compact disk or DVD for business and home use, both in compressed and uncompressed form.

- *Video conferencing*: One of the earliest applications for video compression, video conferencing facilitates meetings between participants in two or more separate locations.

- *Video telephony*: Often used interchangeably with video conferencing, this usually means a face-to-face discussion between two parties via a video 'link'.

- *Remote learning*: There is an increasing interest in the provision of computer-based learning to supplement or replace traditional 'face-to-face' teaching and learning. Digital video is seen as an important component of this in the form of stored video material and video conferencing.

- *Remote medicine*: Medical support provided at a distance, or 'telemedicine', is another potential growth area where digital video and images may be used together with other monitoring techniques to provide medical advice at a distance.

- *Television*: Digital television is now widely available and many countries have a time-table for 'switching off' the existing analogue television service. Digital TV is one of the most important mass-market applications for video coding and compression.

- *Video production*: Fully digital video storage, editing and production have been widely used in television studios for many years. The requirement for high image fidelity often means that the popular 'lossy' compression methods described in this book are not an option.

- *Games and entertainment*: The potential for 'real' video imagery in the computer gaming market is just beginning to be realised with the convergence of 3-D graphics and 'natural' video.

### 2.7.1  Platforms

Developers are targeting an increasing range of platforms to run the ever-expanding list of digital video applications.

*Dedicated platforms* are designed to support a specific video application and no other. Examples include digital video cameras, dedicated video conferencing systems, digital TV set-top boxes and DVD players. In the early days, the high processing demands of digital video meant that dedicated platforms were the only practical design solution. Dedicated platforms will continue to be important for low-cost, mass-market systems but are increasingly being replaced by more flexible solutions.

The *PC* has emerged as a key platform for digital video. A continual increase in PC processing capabilities (aided by hardware enhancements for media applications such as the Intel MMX instructions) means that it is now possible to support a wide range of video applications from video editing to real-time video conferencing.

*Embedded platforms* are an important new market for digital video techniques. For example, the personal communications market is now huge, driven mainly by users of mobile telephones. Video services for mobile devices (running on low-cost embedded processors) are seen as a major potential growth area. This type of platform poses many challenges for application developers due to the limited processing power, relatively poor wireless communications channel and the requirement to keep equipment and usage costs to a minimum.

## 2.8  SUMMARY

Sampling of an analogue video signal, both spatially and temporally, produces a digital video signal. Representing a colour scene requires at least three separate 'components': popular colour 'spaces' include red/green/blue and Y/Cr/Cb (which has the advantage that the chrominance may be subsampled to reduce the information rate without significant loss of quality). The human observer's response to visual information affects the way we perceive video quality and this is notoriously difficult to quantify accurately. Subjective tests (involving 'real' observers) are time-consuming and expensive to run; objective tests range from the simplistic (but widely used) PSNR measure to complex models of the human visual system.

The digital video applications listed above have been made possible by the development of compression or coding technology. In the next chapter we introduce the basic concepts of video and image compression.

# REFERENCES

1. Recommendation ITU-T BT.500-10, 'Methodology for the subjective assessment of the quality of television pictures', ITU-T, 2000.
2. R. Aldridge, J. Davidoff, M. Ghanbari, D. Hands and D. Pearson, 'Subjective assessment of time-varying coding distortions', *Proc. PCS96*, Melbourne, March 1996.
3. C. J. van den Branden Lambrecht and O. Verscheure, 'Perceptual quality measure using a spatio-temporal model of the Human Visual System', *Digital Video Compression Algorithms and Technologies, Proc. SPIE*, Vol. 2668, San Jose, 1996.
4. H. Wu, Z. Yu, S. Winkler and T. Chen, 'Impairment metrics for MC/DPCM/DCT encoded digital video', *Proc. PCS01*, Seoul, April 2001.
5. K. T. Tan and M. Ghanbari, 'A multi-metric objective picture quality measurement model for MPEG video', *IEEE Trans. CSVT*, **10**(7), October 2000.
6. A. Basso, I. Dalgiç, F. Tobagi and C. J. van den Branden Lambrecht, 'A feedback control scheme for low latency constant quality MPEG-2 video encoding', *Digital Compression Technologies and Systems for Video Communications, Proc. SPIE*, Vol. 2952, Berlin, 1996.
7. http://www.vqeg.org/ [Video Quality Experts Group].
8. Recommendation ITU-R BT.601-5, 'Studio encoding parameters of digital television for standard 4 : 3 and wide-screen 16 : 9 aspect ratios', ITU-T, 1995.

# 3

# Image and Video Compression Fundamentals

## 3.1 INTRODUCTION

Representing video material in a digital form requires a large number of bits. The volume of data generated by digitising a video signal is too large for most storage and transmission systems (despite the continual increase in storage capacity and transmission 'bandwidth'). This means that compression is essential for most digital video applications.

The ITU-R 601 standard (described in Chapter 2) describes a digital format for video that is roughly equivalent to analogue television, in terms of spatial resolution and frame rate. One channel of ITU-R 601 television, broadcast in uncompressed digital form, requires a transmission bit rate of 216 Mbps. At this bit rate, a 4.7 Gbyte DVD could store just 87 seconds of uncompressed video.

Table 3.1 shows the uncompressed bit rates of several popular video formats. From this table it can be seen that even QCIF at 15 frames per second (i.e. relatively low-quality video, suitable for video telephony) requires 4.6 Mbps for transmission or storage. Table 3.2 lists typical capacities of popular storage media and transmission networks.

There is a clear gap between the high bit rates demanded by uncompressed video and the available capacity of current networks and storage media. The purpose of video compression (video coding) is to fill this gap. A video compression system aims to reduce the amount of data required to store or transmit video whilst maintaining an 'acceptable' level of video quality. Most of the practical systems and standards for video compression are 'lossy', i.e. the volume of data is reduced (compressed) at the expense of a loss of visual quality. The quality loss depends on many factors, but in general, higher compression results in a greater loss of quality.

### 3.1.1 Do We Need Compression?

The following statement (or something similar) has been made many times over the 20-year history of image and video compression: 'Video compression will become redundant very soon, once transmission and storage capacities have increased to a sufficient level to cope with uncompressed video.' It is true that both storage and transmission capacities continue to increase. However, an efficient and well-designed video compression system gives very significant performance advantages for visual communications at both low and high transmission bandwidths. At low bandwidths, compression enables applications that would not otherwise be possible, such as basic-quality video telephony over a standard telephone

**Table 3.1**  Uncompressed bit rates

| Format | Luminance resolution | Chrominance resolution | Frames per second | Bits per second (uncompressed) |
|---|---|---|---|---|
| ITU-R 601 | $858 \times 525$ | $429 \times 525$ | 30 | 216 Mbps |
| CIF | $352 \times 288$ | $176 \times 144$ | 30 | 36.5 Mbps |
| QCIF | $176 \times 144$ | $88 \times 72$ | 15 | 4.6 Mbps |

**Table 3.2**  Typical transmission / storage capacities

| Media / network | Capacity |
|---|---|
| Ethernet LAN (10 Mbps) | Max. 10 Mbps / Typical 1–2 Mbps |
| ADSL | Typical 1–2 Mbps (downstream) |
| ISDN-2 | 128 kbps |
| V.90 modem | 56 kbps downstream / 33 kbps upstream |
| DVD-5 | 4.7 Gbytes |
| CD-ROM | 640 Mbytes |

connection. At high bandwidths, compression can support a much higher visual quality. For example, a 4.7 Gbyte DVD can store approximately 2 hours of uncompressed QCIF video (at 15 frames per second) or 2 hours of compressed ITU-R 601 video (at 30 frames per second). Most users would prefer to see 'television-quality' video with smooth motion rather than 'postage-stamp' video with jerky motion.

Video compression and video CODECs will therefore remain a vital part of the emerging multimedia industry for the foreseeable future, allowing designers to make the most efficient use of available transmission or storage capacity. In this chapter we introduce the basic components of an image or video compression system. We begin by defining the concept of an image or video encoder (compressor) and decoder (decompressor). We then describe the main functional blocks of an image encoder/decoder (CODEC) and a video CODEC.

## 3.2  IMAGE AND VIDEO COMPRESSION

Information-carrying signals may be *compressed*, i.e. converted to a representation or form that requires fewer bits than the original (uncompressed) signal. A device or program that compresses a signal is an *encoder* and a device or program that decompresses a signal is a *decoder*. An enCOder/DECoder pair is a *CODEC*.

Figure 3.1 shows a typical example of a CODEC as part of a communication system. The original (uncompressed) information is encoded (compressed): this is *source coding*. The source coded signal is then encoded further to add error protection (*channel coding*) prior to transmission over a *channel*. At the receiver, a *channel decoder* detects and/or corrects transmission errors and a *source decoder* decompresses the signal. The decompressed signal may be identical to the original signal (*lossless compression*) or it may be distorted or degraded in some way (*lossy compression*).

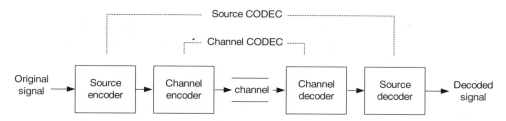

**Figure 3.1** Source coder, channel coder, channel

General-purpose compression CODECs are available that are designed to encode and compress data containing *statistical redundancy*. An information-carrying signal usually contains redundancy, which means that it may (in theory) be represented in a more compact way. For example, characters within a text file occur with varying frequencies: in English, the letters E, T and A occur more often than the letters Q, Z and X. This makes it possible to compress a text file by representing frequently occurring characters with short codes and infrequently occurring characters with longer codes (this principle is used in *Huffman coding*, described in Chapter 8). Compression is achieved by reducing the statistical redundancy in the text file. This type of general-purpose CODEC is known as an *entropy CODEC*.

Photographic images and sequences of video frames are not amenable to compression using general-purpose CODECs. Their contents (pixel values) tend to be highly correlated, i.e. neighbouring pixels have similar values, whereas an entropy encoder performs best with data values that have a certain degree of independence (decorrelated data). Figure 3.2 illustrates the poor performance of a general-purpose entropy encoder with image data. The original image (a) is compressed and decompressed using a ZIP program to produce

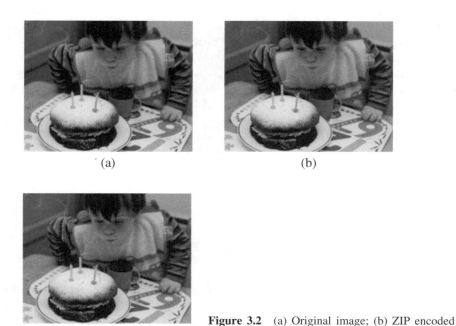

(a)

(b)

(c)

**Figure 3.2** (a) Original image; (b) ZIP encoded and decoded; (c) JPEG encoded and decoded

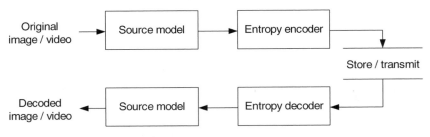

**Figure 3.3**   Image or video CODEC

image (b). This is identical to the original (lossless compression), but the compressed file is only 92% of the size of the original, i.e. there is very little compression. Image (c) is obtained by compressing and decompressing the original using the JPEG compression method. The compressed version is less than a quarter of the size of the original (over $4 \times$ compression) and the decompressed image looks almost identical to the original. (It is in fact slightly 'degraded' due to the lossy compression process.)

In this example, the JPEG method achieved good compression performance by applying a *source model* to the image before compression. The source model attempts to exploit the properties of video or image data and to represent it in a form that can readily be compressed by an entropy encoder. Figure 3.3 shows the basic design of an image or video CODEC consisting of a source model and an entropy encoder/decoder.

Images and video signals have a number of properties that may be exploited by source models. Neighbouring samples (pixels) within an image or a video frame tend to be highly correlated and so there is significant spatial redundancy. Neighbouring regions within successive video frames also tend to be highly correlated (temporal redundancy). As well as these statistical properties (statistical redundancy), a source model may take advantage of *subjective redundancy*, exploiting the sensitivity of the human visual system to various characteristics of images and video. For example, the HVS is much more sensitive to low frequencies than to high ones and so it is possible to compress an image by eliminating certain high-frequency components. Image (c) in Figure 3.2 was compressed by discarding certain subjectively redundant components of the information: the decoded image is not identical to the original but the information loss is not obvious to the human viewer. Examples of image and video source models include the following:

### 3.2.1   DPCM (Differential Pulse Code Modulation)

Each sample or pixel is predicted from one or more previously transmitted samples. The simplest prediction is formed from the previous pixel (pixel A in Figure 3.4). A more accurate prediction can be obtained using a weighted average of neighbouring pixels (for example, A, B and C in Figure 3.4). The actual pixel value $X$ is subtracted from the prediction and the difference (the prediction error) is transmitted to the receiver. The prediction error will typically be small due to spatial correlation, and compression can be achieved by representing common, small prediction errors with short binary codes and larger, less common errors with longer codes. Further compression may be achieved by quantising the prediction error and reducing its precision: this is lossy compression as it

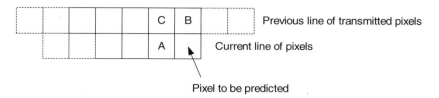

Previous line of transmitted pixels

Current line of pixels

Pixel to be predicted

**Figure 3.4**  DPCM

becomes impossible to exactly reproduce the original values at the decoder. DPCM may be applied spatially (using adjacent pixels in the same frame) and/or temporally (using adjacent pixels in a previous frame to form the prediction) and gives modest compression with low complexity.

### 3.2.2  Transform Coding

The image samples are transformed into another domain (or representation) and are represented by transform *coefficients*. In the 'spatial domain' (i.e. the original form of the image), samples are highly spatially correlated. The aim of transform coding is to reduce this correlation, ideally leaving a small number of visually significant transform coefficients (important to the appearance of the original image) and a large number of insignificant coefficients (that may be discarded without significantly affecting the visual quality of the image). The transform process itself does not achieve compression: a lossy quantisation process in which the insignificant coefficients are removed, leaving behind a small number of significant coefficients, usually follows it. Transform coding (Figure 3.5) forms the basis of most of the popular image and video compression systems and is described in more detail in this chapter and in Chapter 7.

### 3.2.3  Motion-compensated Prediction

Using a similar principle to DPCM, the encoder forms a model of the current frame based on the samples of a previously transmitted frame. The encoder attempts to 'compensate' for motion in a video sequence by translating (moving) or warping the samples of the previously transmitted 'reference' frame. The resulting *motion-compensated* predicted frame (the model of the current frame) is subtracted from the current frame to produce a residual 'error' frame (Figure 3.6). Further coding usually follows motion-compensated prediction, e.g. transform coding of the residual frame.

**Figure 3.5**  Transform coding

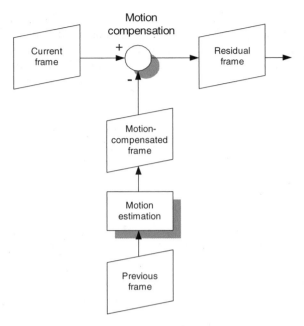

**Figure 3.6**   Motion-compensated prediction

## 3.2.4   Model-based Coding

The encoder attempts to create a semantic model of the video scene, for example by analysing and interpreting the content of the scene. An example is a 'talking head' model: the encoder analyses a scene containing a person's head and shoulders (a typical video conferencing scene) and models the head as a 3-D object. The decoder maintains its own 3-D model of the head. Instead of transmitting information that describes the entire image, the encoder sends only the animation parameters required to 'move' the model, together with an error signal that compensates for the difference between the modelled scene and the actual video scene (Figure 3.7). Model-based coding has the potential for far greater

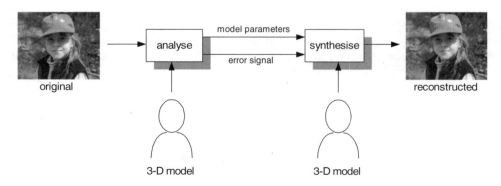

**Figure 3.7**   Model-based coding

compression than the other source models described above: however, the computational complexity required to analyse and synthesise 3-D models of a video scene in real time is very high.

## 3.3 IMAGE CODEC

An image CODEC encodes and decodes single images or individual frames from a video sequence (Figure 3.8) and may consist of a *transform coding* stage followed by *quantisation* and *entropy coding*.

### 3.3.1 Transform Coding

The transform coding stage converts (transforms) the image from the spatial domain into another domain in order to make it more amenable to compression. The transform may be applied to discrete blocks of the image (block transform) or to the entire image.

#### *Block transforms*

The spatial image samples are processed in discrete blocks, typically $8 \times 8$ or $16 \times 16$ samples. Each block is transformed using a 2-D transform to produce a block of transform coefficients. The performance of a block-based transform for image compression depends on how well it can decorrelate the information in each block.

The Karhunen–Loeve transform (KLT) has the 'best' performance of any block-based image transform. The coefficients produced by the KLT are decorrelated and the energy in the block is packed into a minimal number of coefficients. The KLT is, however, very computationally inefficient, and it is impractical because the functions required to carry out the transform ('basis functions') must be calculated in advance and transmitted to the decoder for every image. The discrete cosine transform (DCT) performs nearly as well as the KLT and is much more computationally efficient. Figure 3.9 shows a $16 \times 16$ block of

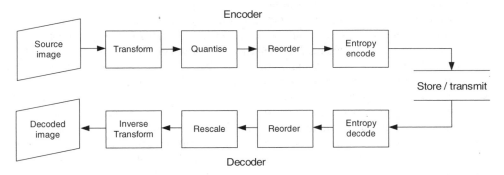

**Figure 3.8** Image CODEC

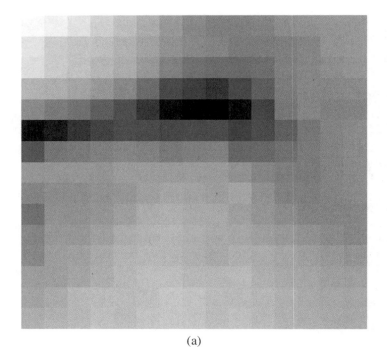

(a)

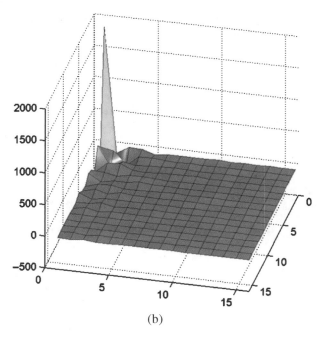

(b)

**Figure 3.9**    (a) 16 × 16 block of pixels; (b) DCT coefficients

image samples (a) and the corresponding block of coefficients produced by the DCT (b). In the original block, the energy is distributed across the 256 samples and the latter are clearly closely interrelated (correlated). In the coefficient block, the energy is concentrated into a few significant coefficients (at the top left). The coefficients are decorrelated: this means that the smaller-valued coefficients may be discarded (for example by quantisation) without significantly affecting the quality of the reconstructed image block at the decoder.

The 16 × 16 array of coefficients shown in Figure 3.9 represent *spatial frequencies* in the original block. At the top left of the array are the low-frequency components, representing the gradual changes of brightness (luminance) in the original block. At the bottom right of the array are high-frequency components and these represent rapid changes in brightness. These frequency components are analogous to the components produced by Fourier analysis of a time-varying signal (and in fact the DCT is closely related to the discrete Fourier transform) except that here the components are 2-D. The example shown in Figure 3.9 is typical for a photographic image: most of the coefficients produced by the DCT are insignificant and can be discarded. This makes the DCT a powerful tool for image and video compression.

### Image transforms

The DCT is usually applied to small, discrete blocks of an image, for reasons of practicality. In contrast, an *image transform* may be applied to a complete video image (or to a large 'tile' within the image). The most popular transform of this type is the *discrete wavelet transform*. A 2-D wavelet transform is applied to the original image in order to decompose it into a series of filtered 'sub-band' images (Figure 3.10). Image (a) is processed in a series of stages to produce the 'wavelet decomposition' image (b). This is made up of a series of components, each containing a subset of the spatial frequencies in the image. At the top left is a low-pass filtered version of the original and moving to the bottom right, each component contains progressively higher-frequency information that adds the 'detail' of the image. It is clear that the higher-frequency components are relatively 'sparse', i.e. many of the values (or 'coefficients') in these components are zero or insignificant. The wavelet transform is thus an efficient way of decorrelating or concentrating the important information into a few significant coefficients.

The wavelet transform is particularly effective for still image compression and has been adopted as part of the JPEG-2000 standard and for still image 'texture' coding in the MPEG-4 standard. Wavelet-based compression is discussed further in Chapter 7.

Another image transform that has received much attention is the so-called *fractal transform*. A fractal transform coder attempts to represent an image as a set of scaled and translated arbitrary 'basis patterns'. Fractal-based coding has not, however, shown sufficiently good performance to be included in any of the international standards for video and image coding and so we will not discuss it in detail.

### 3.3.2 Quantisation

The block and image transforms described above do not themselves achieve any compression. Instead, they represent the image in a different domain in which the image data is

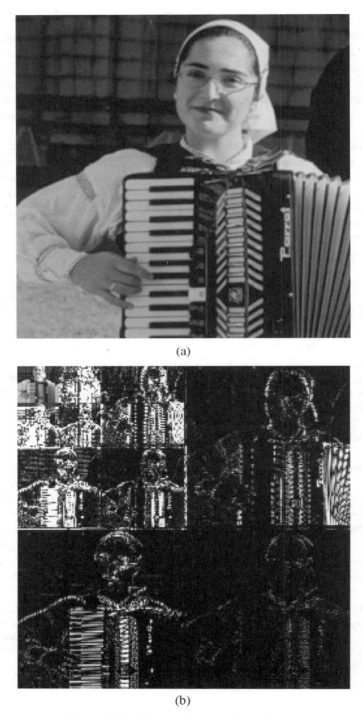

(a)

(b)

**Figure 3.10**    Wavelet decomposition of image

separated into components of varying 'importance' to the appearance of the image. The purpose of quantisation is to remove the components of the transformed data that are unimportant to the visual appearance of the image and to retain the visually important components. Once removed, the less important components cannot be replaced and so quantisation is a lossy process.

*Example*

1. The DCT coefficients shown earlier in Figure 3.9 are quantised by dividing each coefficient by an integer. The resulting array of quantised coefficients is shown in Figure 3.11(a): the large-value coefficients map to non-zero integers and the small-value coefficients map to zero.

2. Rescaling the quantised array (multiplying each coefficient by the same integer) gives Figure 3.11(b). The magnitudes of the larger coefficients are similar to the original coefficients; however, the smaller coefficients (set to zero during quantisation) cannot be recreated and remain at zero.

3. Applying an inverse DCT to the rescaled array gives the block of image samples shown in Figure 3.12: this looks superficially similar to the original image block but some of the information has been lost through quantisation.

It is possible to vary the 'coarseness' of the quantisation process (using a quantiser 'scale factor' or 'step size'). 'Coarse' quantisation will tend to discard most of the coefficients, leaving only the most significant, whereas 'fine' quantisation will tend to leave more coefficients in the quantised block. Coarse quantisation usually gives higher compression at the expense of a greater loss in image quality. The quantiser scale factor or step size is often the main parameter used to control image quality and compression in an image or video CODEC. Figure 3.13 shows a small original image (left) and the effect of compression and decompression with fine quantisation (middle) and coarse quantisation (right).

### 3.3.3   Entropy Coding

A typical image block will contain a few significant non-zero coefficients and a large number of zero coefficients after block transform coding and quantisation. The remaining non-zero data can be efficiently compressed using a statistical compression method ('entropy coding'):

1. *Reorder the quantised coefficients*. The non-zero quantised coefficients of a typical image block tend to be clustered around the 'top-left corner', i.e. around the low frequencies (e.g. Figure 3.9). These non-zero values can be grouped together in sequence by reordering the 64 coefficients, for example in a zigzag scanning order (Figure 3.14). Scanning through in a zigzag sequence from the top-left (lowest frequency) to the bottom-right (highest frequency) coefficients groups together the significant low-frequency coefficients.

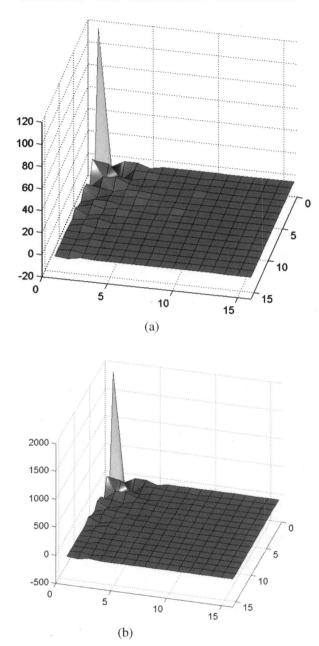

**Figure 3.11**   (a) Quantised DCT coefficients; (b) rescaled

2. Run–level coding. The reordered coefficient array is usually 'sparse', consisting of a group of non-zero coefficients followed by zeros (with occasional non-zero higher-frequency coefficients). This type of array may be compactly represented as a series of (run, level) pairs, as shown in the example in Table 3.3. The first number in the (run,

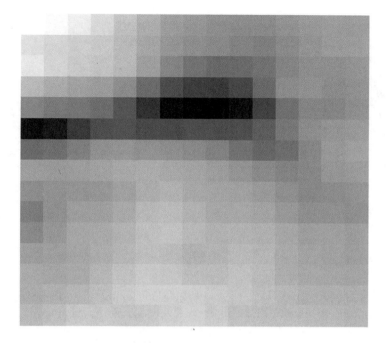

**Figure 3.12** Reconstructed block of image samples

level) pair represents the number of preceding zeros and the second number represents a non-zero value (level). For example, (5, 12) represents five zeros followed by 12.

3. *Entropy coding*. A statistical coding algorithm is applied to the (run, level) data. The purpose of the entropy coding algorithm is to represent frequently occurring (run, level) pairs with a short code and infrequently occurring (run, level) pairs with a longer code. In this way, the run–level data may be compressed into a small number of bits.

Huffman coding and arithmetic coding are widely used for entropy coding of image and video data.

*Huffman coding* replaces each 'symbol' (e.g. a [run, level] pair) with a codeword containing a variable number of bits. The codewords are allocated based on the statistical distribution of

(a)                    (b)                    (c)

**Figure 3.13**   (a) Original image; (b) fine quantisation; (c) coarse quantisation

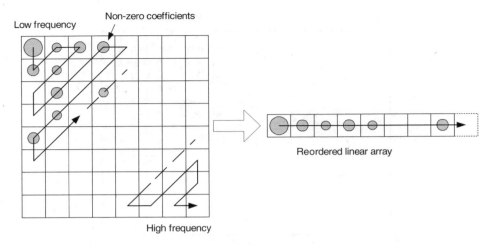

**Figure 3.14**    Zigzag reordering of quantised coefficients

the symbols. Short codewords are allocated to common symbols and longer codewords are allocated to infrequent symbols. Each codeword is chosen to be 'uniquely decodeable', so that a decoder can extract the series of variable-length codewords without ambiguity. Huffman coding is well suited to practical implementation and is widely used in practice.

*Arithmetic coding* maps a series of symbols to a fractional number (see Chapter 8) that is then converted into a binary number and transmitted. Arithmetic coding has the potential for higher compression than Huffman coding. Each symbol may be represented with a fractional number of bits (rather than just an integral number of bits) and this means that the bits allocated per symbol may be more accurately matched to the statistical distribution of the coded data.

### 3.3.4   Decoding

The output of the entropy encoder is a sequence of binary codes representing the original image in compressed form. In order to recreate the image it is necessary to decode this sequence and the decoding process (shown in Figure 3.8) is almost the reverse of the encoding process.
    An entropy decoder extracts run–level symbols from the bit sequence. These are converted to a sequence of coefficients that are reordered into a block of quantised coefficients. The decoding operations up to this point are the inverse of the equivalent encoding operations. Each coefficient is multiplied by the integer scale factor ('rescaled'). This is often described

**Table 3.3**   Run–level coding example

| | |
|---|---|
| Reordered coefficient data | 24, 3, $-9$, 0, $-2$, 0, 0, 0, 0, 0, 12, 0, 0, 0, 2,... |
| Run–level pairs | (0, 24), (0, 3), (0, $-9$), (1, $-2$), (5, 12), (3, 2)... |

as 'inverse quantisation', but in fact the loss of precision due to quantisation cannot be reversed and so the rescaled coefficients are not identical to the original transform coefficients.

The rescaled coefficients are transformed with an inverse transform to reconstruct a decoded image. Because of the data loss during quantisation, this image will not be identical to the original image: the amount of difference depends partly on the 'coarseness' of quantisation.

## 3.4 VIDEO CODEC

A video signal consists of a sequence of individual frames. Each frame may be compressed individually using an image CODEC as described above: this is described as *intra-frame coding*, where each frame is 'intra' coded without any reference to other frames. However, better compression performance may be achieved by exploiting the temporal redundancy in a video sequence (the similarities between successive video frames). This may be achieved by adding a 'front end' to the image CODEC, with two main functions:

1. *Prediction*: create a prediction of the current frame based on one or more previously transmitted frames.

2. *Compensation*: subtract the prediction from the current frame to produce a 'residual frame'.

The residual frame is then processed using an 'image CODEC'. The key to this approach is the prediction function: if the prediction is accurate, the residual frame will contain little data and will hence be compressed to a very small size by the image CODEC. In order to decode the frame, the decoder must 'reverse' the compensation process, adding the prediction to the decoded residual frame (*reconstruction*) (Figure 3.15). This is *inter-frame coding*: frames are coded based on some relationship with other video frames, i.e. coding exploits the interdependencies of video frames.

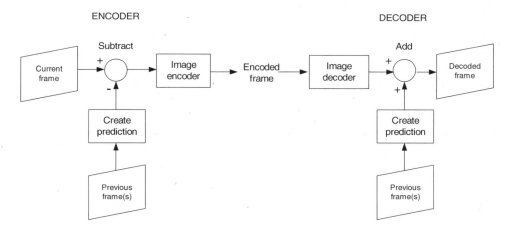

**Figure 3.15**   Video CODEC with prediction

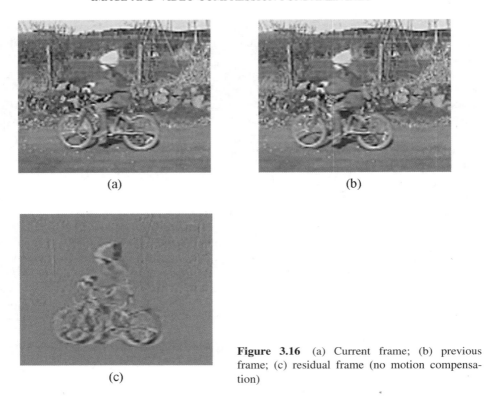

(a)

(b)

(c)

**Figure 3.16** (a) Current frame; (b) previous frame; (c) residual frame (no motion compensation)

## 3.4.1  Frame Differencing

The simplest predictor is just the previous transmitted frame. Figure 3.16 shows the residual frame produced by subtracting the previous frame from the current frame in a video sequence. Mid-grey areas of the residual frame contain zero data: light and dark areas indicate positive and negative residual data respectively. It is clear that much of the residual data is zero: hence, compression efficiency can be improved by compressing the residual frame rather than the current frame.

**Table 3.4** Prediction 'drift'

| Encoder input | Encoder prediction | Encoder output/ decoder input | Decoder prediction | Decoder output |
|---|---|---|---|---|
| Original frame 1 | Zero | Compressed frame 1 | Zero | Decoded frame 1 |
| Original frame 2 | Original frame 1 | Compressed residual frame 2 | Decoded frame 1 | Decoded frame 2 |
| Original frame 3 | Original frame 2 | Compressed residual frame 2 | Decoded frame 2 | Decoded frame 3 |
| . . . | | | | . . . |

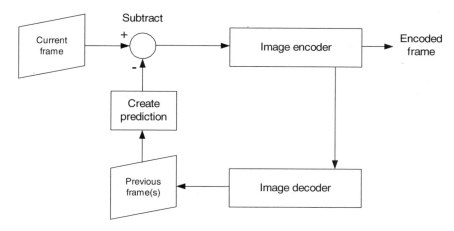

**Figure 3.17**  Encoder with decoding loop

The decoder faces a potential problem that can be illustrated as follows. Table 3.4 shows the sequence of operations required to encode and decode a series of video frames using frame differencing. For the first frame the encoder and decoder use no prediction. The problem starts with frame 2: the encoder uses the original frame 1 as a prediction and encodes the resulting residual. However, the decoder only has the decoded frame 1 available to form the prediction. Because the coding process is lossy, there is a difference between the decoded and original frame 1 which leads to a small error in the prediction of frame 2 at the decoder. This error will build up with each successive frame and the encoder and decoder predictors will rapidly 'drift' apart, leading to a significant drop in decoded quality.

The solution to this problem is for the encoder to use a decoded frame to form the prediction. Hence the encoder in the above example decodes (or reconstructs) frame 1 to form a prediction for frame 2. The encoder and decoder use the same prediction and drift should be reduced or removed. Figure 3.17 shows the complete encoder which now includes a decoding 'loop' in order to reconstruct its prediction reference. The reconstructed (or 'reference') frame is stored in the encoder and in the decoder to form the prediction for the next coded frame.

### 3.4.2  Motion-compensated Prediction

Frame differencing gives better compression performance than intra-frame coding when successive frames are very similar, but does not perform well when there is a significant change between the previous and current frames. Such changes are usually due to movement in the video scene and a significantly better prediction can be achieved by *estimating* this movement and *compensating* for it.

Figure 3.18 shows a video CODEC that uses motion-compensated prediction. Two new steps are required in the encoder:

1. Motion estimation: a region of the current frame (often a rectangular block of luminance samples) is compared with neighbouring regions of the previous reconstructed frame.

ENCODER                                        DECODER

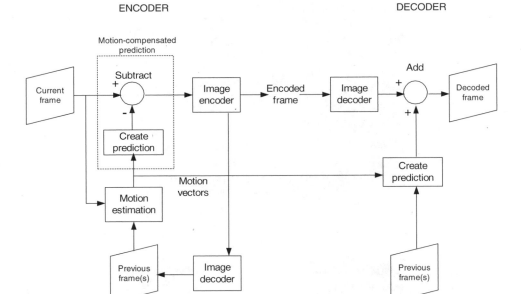

**Figure 3.18**   Video CODEC with motion estimation and compensation

The motion estimator attempts to find the 'best match', i.e. the neighbouring block in the reference frame that gives the smallest residual block.

2. Motion compensation: the 'matching' region or block from the reference frame (identified by the motion estimator) is subtracted from the current region or block.

The decoder carries out the same motion compensation operation to reconstruct the current frame. This means that the encoder has to transmit the location of the 'best' matching blocks to the decoder (typically in the form of a set of *motion vectors*).

Figure 3.19 shows a residual frame produced by subtracting a motion-compensated version of the previous frame from the current frame (shown in Figure 3.16). The residual frame clearly contains less data than the residual in Figure 3.16. This improvement in

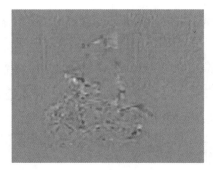

**Figure 3.19**   Residual frame (MCP)

compression does not come without a price: motion estimation can be very computationally intensive. The design of a motion estimation algorithm can have a dramatic effect on
the compression performance and computational complexity of a video CODEC.

### 3.4.3   Transform, Quantisation and Entropy Encoding

A block or image transform is applied to the residual frame and the coefficients are quantised
and reordered. Run–level pairs are entropy coded as before (although the statistical
distribution and hence the coding tables are generally different for inter-coded data). If
motion-compensated prediction is used, motion vector information must be sent in addition
to the run–level data. The motion vectors are typically entropy encoded in a similar way to
run–level pairs, i.e. commonly occurring motion vectors are coded with shorter codes and
uncommon vectors are coded with longer codes.

### 3.4.4   Decoding

A motion-compensated decoder (Figure 3.18) is usually simpler than the corresponding
encoder. The decoder does not need a motion estimation function (since the motion
information is transmitted in the coded bit stream) and it contains only a decoding path
(compared with the encoding and decoding paths in the encoder).

## 3.5   SUMMARY

Efficient coding of images and video sequences involves creating a model of the source data
that converts it into a form that can be compressed. Most image and video CODECs
developed over the last two decades have been based around a common set of 'building
blocks'. For motion video compression, the first step is to create a motion-compensated
prediction of the frame to be compressed, based on one or more previously transmitted
frames. The difference between this model and the actual input frame is then coded using an
image CODEC. The data is transformed into another domain (e.g. the DCT or wavelet
domain), quantised, reordered and compressed using an entropy encoder. A decoder must
reverse these steps to reconstruct the frame: however, quantisation cannot be reversed and so
the decoded frame is an imperfect copy of the original.

   An encoder and decoder must clearly use a compatible set of algorithms in order to
successfully exchange compressed image or video data. Of prime importance is the syntax or
structure of the compressed data. In the past 15 years there has been a significant worldwide
effort to develop standards for video and image compression. These standards generally
describe a syntax (and a decoding process) to support video or image communications for a
wide range of applications. Chapters 4 and 5 provide an overview of the main standards
bodies and JPEG, MPEG and H.26x video and image coding standards.

# 4

# Video Coding Standards: JPEG and MPEG

## 4.1  INTRODUCTION

The majority of video CODECs in use today conform to one of the international standards for video coding. Two standards bodies, the International Standards Organisation (ISO) and the International Telecommunications Union (ITU), have developed a series of standards that have shaped the development of the visual communications industry. The ISO JPEG and MPEG-2 standards have perhaps had the biggest impact: JPEG has become one of the most widely used formats for still image storage and MPEG-2 forms the heart of digital television and DVD-video systems. The ITU's H.261 standard was originally developed for video conferencing over the ISDN, but H.261 and H.263 (its successor) are now widely used for real-time video communications over a range of networks including the Internet.

This chapter begins by describing the process by which these standards are proposed, developed and published. We describe the popular ISO coding standards, JPEG and JPEG-2000 for still images, MPEG-1, MPEG-2 and MPEG-4 for moving video. In Chapter 5 we introduce the ITU-T H.261, H.263 and H.26L standards.

## 4.2  THE INTERNATIONAL STANDARDS BODIES

It was recognised in the 1980s that video coding and transmission could become a commercially important application area. The development of video coding technology since then has been bound up with a series of international standards for image and video coding. Each of these standards supports a particular application of video coding (or a set of applications), such as video conferencing and digital television. The aim of an image or video coding standard is to support a particular class of application and to encourage interoperability between equipment and systems from different manufacturers. Each standard describes a *syntax* or method of representation for compressed images or video. The developers of each standard have attempted to incorporate the best developments in video coding technology (in terms of coding efficiency and ease of practical implementation).

Each of the international standards takes a similar approach to meeting these goals. A video coding standard describes syntax for representing compressed video data and the procedure for decoding this data as well as (possibly) a 'reference' decoder and methods of proving conformance with the standard.

In order to provide the maximum flexibility and scope for innovation, the standards do not define a video or image encoder: this is left to the designer's discretion. However, in practice the syntax elements and reference decoder limit the scope for alternative designs that still meet the requirements of the standard.

## 4.2.1 The Expert Groups

The most important developments in video coding standards have been due to two international standards bodies: the ITU (formerly the CCITT)[1] and the ISO.[2] The ITU has concentrated on standards to support real-time, two-way video communications. The group responsible for developing these standards is known as VCEG (Video Coding Experts Group) and has issued:

- H.261 (1990): Video telephony over constant bit-rate channels, primarily aimed at ISDN channels of $p \times 64$ kbps.

- H.263 (1995): Video telephony over circuit- and packet-switched networks, supporting a range of channels from low bit rates (20–30 kbps) to high bit rates (several Mbps).

- H.263+ (1998), H.263++ (2001): Extensions to H.263 to support a wider range of transmission scenarios and improved compression performance.

- H.26L (under development): Video communications over channels ranging from very low (under 20 kbps) to high bit rates.

The H.26$x$ series of standards will be described in Chapter 5. In parallel with the ITU's activities, the ISO has issued standards to support storage and distribution applications. The two relevant groups are JPEG (Joint Photographic Experts Group) and MPEG (Moving Picture Experts Group) and they have been responsible for:

- JPEG (1992)[3]: Compression of still images for storage purposes.

- MPEG-1 (1993)[4]: Compression of video and audio for storage and real-time play back on CD-ROM (at a bit rate of 1.4 Mbps).

- MPEG-2 (1995)[5]: Compression and transmission of video and audio programmes for storage and broadcast applications (at typical bit rates of 3–5 Mbps and above).

- MPEG-4 (1998)[6]: Video and audio compression and transport for multimedia terminals (supporting a wide range of bit rates from around 20–30 kbps to high bit rates).

- JPEG-2000 (2000)[7]: Compression of still images (featuring better compression performance than the original JPEG standard).

Since releasing Version 1 of MPEG-4, the MPEG committee has concentrated on 'framework' standards that are not primarily concerned with video coding:

- MPEG-7[8]: Multimedia Content Description Interface. This is a standard for describing multimedia content data, with the aim of providing a standardised system for content-based indexing and retrieval of multimedia information. MPEG-7 is concerned with access to multimedia data rather than the mechanisms for coding and compression. MPEG-7 is scheduled to become an international standard in late 2001.

- MPEG-21[9]: Multimedia Framework. The MPEG-21 initiative looks beyond coding and indexing to the complete multimedia content 'delivery chain', from creation through production and delivery to 'consumption' (e.g. viewing the content). MPEG-21 will define key elements of this delivery framework, including content description and identification, content handling, intellectual property management, terminal and network interoperation and content representation. The motivation behind MPEG-21 is to encourage integration and interoperation between the diverse technologies that are required to create, deliver and decode multimedia data. Work on the proposed standard started in June 2000.

Figure 4.1 shows the relationship between the standards bodies, the expert groups and the video coding standards. The expert groups have addressed different application areas (still images, video conferencing, entertainment and multimedia), but in practice there are many overlaps between the applications of the standards. For example, a version of JPEG, Motion JPEG, is widely used for video conferencing and video surveillance; MPEG-1 and MPEG-2 have been used for video conferencing applications; and the core algorithms of MPEG-4 and H.263 are identical.

In recognition of these natural overlaps, the expert groups have cooperated at several stages and the result of this cooperation has led to outcomes such as the ratification of MPEG-2 (Video) as ITU standard H.262 and the incorporation of 'baseline' H.263 into MPEG-4 (Video). There is also interworking between the VCEG and MPEG committees and

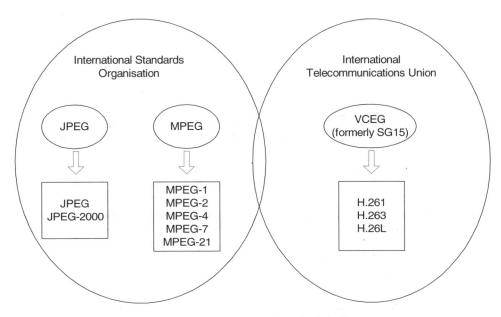

**Figure 4.1**   International standards bodies

other related bodies such as the Internet Engineering Task Force (IETF), industry groups (such as the Digital Audio Visual Interoperability Council, DAVIC) and other groups within ITU and ISO.

## 4.2.2 The Standardisation Process

The development of an international standard for image or video coding is typically an involved process:

1. The scope and aims of the standard are defined. For example, the emerging H.26L standard is designed with real-time video communications applications in mind and aims to improve performance over the preceding H.263 standard.

2. Potential technologies for meeting these aims are evaluated, typically by competitive testing. The test scenario and criteria are defined and interested parties are encouraged to participate and demonstrate the performance of their proposed solutions. The 'best' technology is chosen based on criteria such as coding performance and implementation complexity.

3. The chosen technology is implemented as a *test model*. This is usually a software implementation that is made available to members of the expert group for experimentation, together with a *test model document* that describes its operation.

4. The test model is developed further: improvements and features are proposed and demonstrated by members of the expert group and the best of these developments are integrated into the test model.

5. At a certain point (depending on the timescales of the standardisation effort and on whether the aims of the standard have been sufficiently met by the test model), the model is 'frozen' and the test model document forms the basis of a *draft standard*.

6. The draft standard is reviewed and after approval becomes a published *international standard*.

Officially, the standard is not available in the public domain until the final stage of approval and publication. However, because of the fast-moving nature of the video communications industry, draft documents and test models can be very useful for developers and manufacturers. Many of the ITU VCEG documents and models are available via public FTP.[10] Most of the MPEG working documents are restricted to members of MPEG itself, but a number of overview documents are available at the MPEG website.[11] Information and links about JPEG and MPEG are available.[12,13] Keeping in touch with the latest developments and gaining access to draft standards are powerful reasons for companies and organisations to become involved with the MPEG, JPEG and VCEG committees.

## 4.2.3 Understanding and Using the Standards

Published ITU and ISO standards may be purchased from the relevant standards body.[1,2] For developers of standards-compliant video coding systems, the published standard is an essential point of reference as it defines the syntax and capabilities that a video CODEC must conform to in order to successfully interwork with other systems. However, the standards themselves are not an ideal introduction to the concepts and techniques of video coding: the aim of the standard is to define the syntax as explicitly and unambiguously as possible and this does not make for easy reading.

Furthermore, the standards do not necessarily indicate practical constraints that a designer must take into account. Practical issues and good design techniques are deliberately left to the discretion of manufacturers in order to encourage innovation and competition, and so other sources are a much better guide to practical design issues. This book aims to collect together information and guidelines for designers and integrators; other texts that may be useful for developers are listed in the bibliography.

The test models produced by the expert groups are designed to facilitate experimentation and comparison of alternative techniques, and the test model (a software model with an accompanying document) can provide a valuable insight into the implementation of the standard. Further documents such as implementation guides (e.g. H.263 Appendix III[14]) are produced by the expert groups to assist with the interpretation of the standards for practical applications.

In recent years the standards bodies have recognised the need to direct developers towards certain subsets of the tools and options available within the standard. For example, H.263 now has a total of 19 optional modes and it is unlikely that any particular application would need to implement all of these modes. This has led to the concept of *profiles* and *levels*. A 'profile' describes a subset of functionalities that may be suitable for a particular application and a 'level' describes a subset of operating resolutions (such as frame resolution and frame rates) for certain applications.

## 4.3 JPEG (JOINT PHOTOGRAPHIC EXPERTS GROUP)

### 4.3.1 JPEG

International standard ISO 10918[3] is popularly known by the acronym of the group that developed it, the Joint Photographic Experts Group. Released in 1992, it provides a method and syntax for compressing continuous-tone still images (such as photographs). Its main application is storage and transmission of still images in a compressed form, and it is widely used in digital imaging, digital cameras, embedding images in web pages, and many more applications. Whilst aimed at still image compression, JPEG has found some popularity as a simple and effective method of compressing moving images (in the form of Motion JPEG).

The JPEG standard defines a syntax and decoding process for a *baseline CODEC* and this includes a set of features that are designed to suit a wide range of applications. Further optional modes are defined that extend the capabilities of the baseline CODEC.

*The baseline CODEC*

A baseline JPEG CODEC is shown in block diagram form in Figure 4.2. Image data is processed one 8 × 8 block at a time. Colour components or planes (e.g. R, G, B or Y, Cr, Cb)

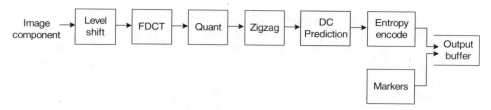

**Figure 4.2**   JPEG baseline CODEC block diagram

may be processed separately (one complete component at a time) or in interleaved order (e.g. a block from each of three colour components in succession). Each block is coded using the following steps.

**Level shift**   Input data is shifted so that it is distributed about zero: e.g. an 8-bit input sample in the range $0:255$ is shifted to the range $-128:127$ by subtracting 128.

**Forward DCT**   An $8 \times 8$ block transform, described in Chapter 7.

**Quantiser**   Each of the 64 DCT coefficients $C_{ij}$ is quantised by integer division:

$$Cq_{ij} = \text{round} \left( \frac{C_{ij}}{Q_{ij}} \right)$$

$Q_{ij}$ is a quantisation parameter and $Cq_{ij}$ is the quantised coefficient. A larger value of $Q_{ij}$ gives higher compression (because more coefficients are set to zero after quantisation) at the expense of increased distortion in the decoded image. The 64 parameters $Q_{ij}$ (one for each coefficient position $ij$) are stored in a quantisation 'map'. The map is not specified by the standard but can be perceptually weighted so that lower-frequency coefficients (DC and low-frequency AC coefficients) are quantised less than higher-frequency coefficients. Figure 4.3

Low frequencies

| 16 | 11 | 10 | 16 | 24 | 40 | 51 | 61 |
|----|----|----|----|----|----|----|----|
| 12 | 12 | 14 | 19 | 26 | 58 | 60 | 55 |
| 14 | 13 | 16 | 24 | 40 | 57 | 69 | 56 |
| 14 | 17 | 22 | 29 | 51 | 87 | 80 | 62 |
| 18 | 22 | 37 | 56 | 68 | 109 | 103 | 77 |
| 24 | 35 | 55 | 64 | 81 | 104 | 113 | 92 |
| 49 | 64 | 78 | 87 | 103 | 121 | 120 | 101 |
| 72 | 92 | 95 | 98 | 112 | 100 | 103 | 99 |

High frequencies    **Figure 4.3**   JPEG quantisation map

gives an example of a quantisation map: the weighting means that the visually important lower frequencies (to the top left of the map) are preserved and the less important higher frequencies (to the bottom right) are more highly compressed.

**Zigzag reordering**    The $8 \times 8$ block of quantised coefficients is rearranged in a zigzag order so that the low frequencies are grouped together at the start of the rearranged array.

**DC differential prediction**    Because there is often a high correlation between the DC coefficients of neighbouring image blocks, a prediction of the DC coefficient is formed from the DC coefficient of the preceding block:

$$DC_{pred} = DC_{cur} - DC_{prev}$$

The prediction $DC_{pred}$ is coded and transmitted, rather than the actual coefficient $DC_{cur}$.

**Entropy encoding**    The differential DC coefficients and AC coefficients are encoded as follows. The number of bits required to represent the DC coefficient, SSSS, is encoded using a variable-length code. For example, SSSS=0 indicates that the DC coefficient is zero; SSSS=1 indicates that the DC coefficient is $+/-1$ (i.e. it can be represented with 1 bit); SSSS=2 indicates that the coefficient is $+3$, $+2$, $-2$ or $-3$ (which can be represented with 2 bits). The actual value of the coefficient, an SSSS-bit number, is appended to the variable-length code (except when SSSS=0).

Each AC coefficient is coded as a variable-length code RRRRSSSS, where RRRR indicates the number of preceding zero coefficients and SSSS indicates the number of bits required to represent the coefficient (SSSS=0 is not required). The actual value is appended to the variable-length code as described above.

*Example*

A run of six zeros followed by the value $+5$ would be coded as:

[RRRR=6] [SSSS=3] [Value=$+5$]

**Marker insertion**    Marker codes are inserted into the entropy-coded data sequence. Examples of markers include the frame header (describing the parameters of the frame such as width, height and number of colour components), scan headers (see below) and restart interval markers (enabling a decoder to resynchronise with the coded sequence if an error occurs).

The result of the encoding process is a compressed sequence of bits, representing the image data, that may be transmitted or stored. In order to view the image, it must be decoded by reversing the above steps, starting with marker detection and entropy decoding and ending with an inverse DCT. Because quantisation is not a reversible process (as discussed in Chapter 3), the decoded image is not identical to the original image.

## Lossless JPEG

JPEG also defines a lossless encoding/decoding algorithm that uses DPCM (described in Chapter 3). Each pixel is predicted from up to three neighbouring pixels and the predicted value is entropy coded and transmitted. Lossless JPEG guarantees image fidelity at the expense of relatively poor compression performance.

## Optional modes

*Progressive encoding* involves encoding the image in a series of progressive 'scans'. The first scan may be decoded to provide a 'coarse' representation of the image; decoding each subsequent scan progressively improves the quality of the image until the final quality is reached. This can be useful when, for example, a compressed image takes a long time to transmit: the decoder can quickly recreate an approximate image which is then further refined in a series of passes. Two versions of progressive encoding are supported: *spectral selection*, where each scan consists of a subset of the DCT coefficients of every block (e.g. (a) DC only; (b) low-frequency AC; (c) high-frequency AC coefficients) and *successive approximation*, where the first scan contains $N$ most significant bits of each coefficient and later scans contain the less significant bits. Figure 4.4 shows an image encoded and decoded using progressive spectral selection. The first image contains the DC coefficients of each block, the second image contains the DC and two lowest AC coefficients and the third contains all 64 coefficients in each block.

(a)

**Figure 4.4**  Progressive encoding example (spectral selection): (a) DC only; (b) DC + two AC; (c) all coefficients

(b)

(c)

**Figure 4.4**   (*Continued*)

*Hierarchical encoding* compresses an image as a series of components at different spatial resolutions. For example, the first component may be a subsampled image at a low spatial resolution, followed by further components at successively higher resolutions. Each successive component is encoded *differentially* from previous components, i.e. only the differences are encoded. A decoder may choose to decode only a subset of the full resolution image; alternatively, the successive components may be used to progressively refine the resolution in a similar way to progressive encoding.

The two progressive encoding modes and the hierarchical encoding mode can be thought of as *scalable coding* modes. Scalable coding will be discussed further in the section on MPEG-2.

## 4.3.2   Motion JPEG

A 'Motion JPEG' or MJPEG CODEC codes a video sequence as a series of JPEG images, each corresponding to one frame of video (i.e. a series of intra-coded frames). Originally, the JPEG standard was not intended to be used in this way: however, MJPEG has become popular and is used in a number of video communications and storage applications. No attempt is made to exploit the inherent temporal redundancy in a moving video sequence and so compression performance is poor compared with inter-frame CODECs (see Chapter 5, 'Performance Comparison'). However, MJPEG has a number of practical advantages:

- *Low complexity*: algorithmic complexity, and requirements for hardware, processing and storage are very low compared with even a basic inter-frame CODEC (e.g. H.261).

- *Error tolerance*: intra-frame coding limits the effect of an error to a single decoded frame and so is inherently resilient to transmission errors. Until recent developments in error resilience (see Chapter 11), MJPEG outperformed inter-frame CODECs in noisy environments.

- *Market awareness*: JPEG is perhaps the most widely known and used of the compression standards and so potential users are already familiar with the technology of Motion JPEG.

Because of its poor compression performance, MJPEG is only suitable for high-bandwidth communications (e.g. over dedicated networks). Perversely, this means that users generally have a good experience of MJPEG because installations do not tend to suffer from the bandwidth and delay problems encountered by inter-frame CODECs used over 'best effort' networks (such as the Internet) or low bit-rate channels. An MJPEG coding integrated circuit(IC), the Zoran ZR36060, is described in Chapter 12.

## 4.3.3   JPEG-2000

The original JPEG standard has gained widespread acceptance and is now ubiquitous throughout computing applications: it is the main format for photographic images on the world wide web and it is widely used for image storage. However, the block-based DCT algorithm has a number of disadvantages, perhaps the most important of which is the 'blockiness' of highly compressed JPEG images (see Chapter 9). Since its release, many alternative coding schemes have been shown to outperform baseline JPEG. The need for better performance at high compression ratios led to the development of the JPEG-2000 standard.[7,15]

The features that JPEG-2000 aims to support are as follows:

- Good compression performance, particularly at high compression ratios.

- Efficient compression of continuous-tone, bi-level and compound images (e.g. photographic images with overlaid text: the original JPEG does not handle this type of image well).

- Lossless and lossy compression (within the same compression framework).

- Progressive transmission (JPEG-2000 supports SNR scalability, a similar concept to JPEG's successive approximation mode, and spatial scalability, similar to JPEG's hierarchical mode).

- Region-of-interest (ROI) coding. This feature allows an encoder to specify an arbitrary region within the image that should be treated differently during encoding: e.g. by encoding the region with a higher quality or by allowing independent decoding of the ROI.

- Error resilience tools including data partitioning (see the description of MPEG-2 below), error detection and concealment (see Chapter 11 for more details).

- Open architecture. The JPEG-2000 standard provides an open 'framework' which should make it relatively easy to add further coding features either as part of the standard or as a proprietary 'add-on' to the standard.

The architecture of a JPEG-2000 encoder is shown in Figure 4.5. This is superficially similar to the JPEG architecture but one important difference is that the same architecture may be used for lossy or lossless coding.

The basic coding unit of JPEG-2000 is a 'tile'. This is normally a $2^n \times 2^n$ region of the image, and the image is 'covered' by non-overlapping identically sized tiles. Each tile is encoded as follows:

- *Transform*: A wavelet transform is carried out on each tile to decompose it into a series of sub-bands (see Sections 3.3.1 and 7.3). The transform may be reversible (for lossless coding applications) or irreversible (suitable for lossy coding applications).

- *Quantisation*: The coefficients of the wavelet transform are quantised (as described in Chapter 3) according to the 'importance' of each sub-band to the final image appearance. There is an option to leave the coefficients unquantised (lossless coding).

- *Entropy coding*: JPEG-2000 uses a form of arithmetic coding to encode the quantised coefficients prior to storage or transmission. Arithmetic coding can provide better compression efficiency than variable-length coding and is described in Chapter 8.

The result is a compression standard that can give significantly better image compression performance than JPEG. For the same image quality, JPEG-2000 can usually compress images by at least twice as much as JPEG. At high compression ratios, the quality of images

**Figure 4.5**  Architecture of JPEG-2000 encoder

degrades gracefully, with the decoded image showing a gradual 'blurring' effect rather than the more obvious blocking effect associated with the DCT. These performance gains are achieved at the expense of increased complexity and storage requirements during encoding and decoding. One effect of this is that images take longer to store and display using JPEG-2000 (though this should be less of an issue as processors continue to get faster).

## 4.4    MPEG (MOVING PICTURE EXPERTS GROUP)

### 4.4.1    MPEG-1

The first standard produced by the Moving Picture Experts Group, popularly known as MPEG-1, was designed to provide video and audio compression for storage and playback on CD-ROMs. A CD-ROM played at 'single speed' has a transfer rate of 1.4 Mbps. MPEG-1 aims to compress video and audio to a bit rate of 1.4 Mbps with a quality that is comparable to VHS videotape. The target market was the 'video CD', a standard CD containing up to 70 minutes of stored video and audio. The video CD was never a commercial success: the quality improvement over VHS tape was not sufficient to tempt consumers to replace their video cassette recorders and the maximum length of 70 minutes created an irritating break in a feature-length movie. However, MPEG-1 is important for two reasons: it has gained widespread use in other video storage and transmission applications (including CD-ROM storage as part of interactive applications and video playback over the Internet), and its functionality is used and extended in the popular MPEG-2 standard.

The MPEG-1 standard consists of three parts. Part 1[16] deals with system issues (including the multiplexing of coded video and audio), Part 2[4] deals with compressed video and Part 3[17] with compressed audio. Part 2 (video) was developed with aim of supporting efficient coding of video for CD playback applications and achieving video quality comparable to, or better than, VHS videotape at CD bit rates (around 1.2 Mbps for video). There was a requirement to minimise decoding complexity since most consumer applications were envisaged to involve decoding and playback only, not encoding. Hence MPEG-1 decoding is considerably simpler than encoding (unlike JPEG, where the encoder and decoder have similar levels of complexity).

*MPEG-1 features*

The input video signal to an MPEG-1 video encoder is 4 : 2 : 0 Y : Cr : Cb format (see Chapter 2) with a typical spatial resolution of $352 \times 288$ or $352 \times 240$ pixels. Each frame of video is processed in units of a *macroblock*, corresponding to a $16 \times 16$ pixel area in the displayed frame. This area is made up of $16 \times 16$ luminance samples, $8 \times 8$ Cr samples and $8 \times 8$ Cb samples (because Cr and Cb have half the horizontal and vertical resolution of the luminance component). A macroblock consists of six $8 \times 8$ blocks: four luminance (Y) blocks, one Cr block and one Cb block (Figure 4.6).

Each frame of video is encoded to produce a coded *picture*. There are three main types: I-pictures, P-pictures and B-pictures. (The standard specifies a fourth picture type, D-pictures, but these are seldom used in practical applications.)

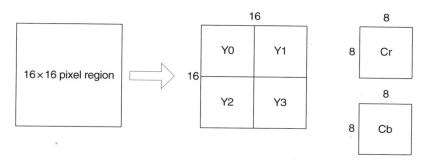

**Figure 4.6**  Structure of a macroblock

*I-pictures* are intra-coded without any motion-compensated prediction (in a similar way to a baseline JPEG image). An I-picture is used as a reference for further predicted pictures (P- and B-pictures, described below).

*P-pictures* are inter-coded using motion-compensated prediction from a *reference picture* (the P-picture or I-picture preceding the current P-picture). Hence a P-picture is predicted using *forward prediction* and a P-picture may itself be used as a reference for further predicted pictures (P- and B-pictures).

*B-pictures* are inter-coded using motion-compensated prediction from *two* reference pictures, the P- and/or I-pictures before and after the current B-picture. Two motion vectors are generated for each macroblock in a B-picture (Figure 4.7): one pointing to a matching area in the previous reference picture (a *forward* vector) and one pointing to a matching area

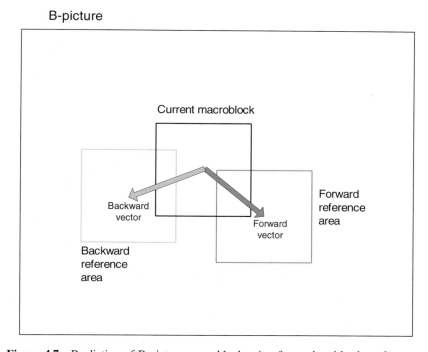

**Figure 4.7**  Prediction of B-picture macroblock using forward and backward vectors

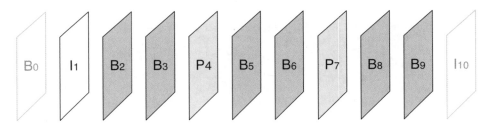

**Figure 4.8** MPEG-1 group of pictures (IBBPBBPBB): display order

in the future reference picture (a *backward* vector). A motion-compensated prediction macroblock can be formed in three ways: forward prediction using the forward vector, backwards prediction using the backward vector or bidirectional prediction (where the prediction reference is formed by averaging the forward and backward prediction references). Typically, an encoder chooses the prediction mode (forward, backward or bidirectional) that gives the lowest energy in the difference macroblock. B-pictures are not themselves used as prediction references for any further predicted frames.

Figure 4.8 shows a typical series of I-, B- and P-pictures. In order to encode a B-picture, two neighbouring I- or P-pictures ('anchor' pictures or 'key' pictures) must be processed and stored in the prediction memory, introducing a delay of several frames into the encoding procedure. Before frame $B_2$ in Figure 4.8 can be encoded, its two 'anchor' frames $I_1$ and $P_4$ must be processed and stored, i.e. frames 1–4 must be processed before frames 2 and 3 can be coded. In this example, there is a delay of at least three frames during encoding (frames 2, 3 and 4 must be stored before $B_2$ can be coded) and this delay will be larger if more B-pictures are used.

In order to limit the delay at the decoder, encoded pictures are *reordered* before transmission, such that all the anchor pictures required to decode a B-picture are placed *before* the B-picture. Figure 4.9 shows the same series of frames, reordered prior to transmission. $P_4$ is now placed *before* $B_2$ and $B_3$. Decoding proceeds as shown in Table 4.1: $P_4$ is decoded immediately after $I_1$ and is stored by the decoder. $B_2$ and $B_3$ can now be decoded and displayed (because their prediction references, $I_1$ and $P_4$, are both available), after which $P_4$ is displayed. There is at most one frame delay between decoding and display and the decoder only needs to store two decoded frames. This is one example of 'asymmetry' between encoder and decoder: the delay and storage in the decoder are significantly lower than in the encoder.

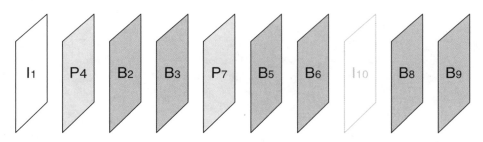

**Figure 4.9** MPEG-1 group of pictures: transmission order

**Table 4.1**   MPEG-1 decoding and display order

| Decode | Display |
| --- | --- |
| $I_1$ | $I_1$ |
| $P_4$ | — |
| $B_2$ | $B_2$ |
| $B_3$ | $B_3$ |
| — | $P_4$ |
| $P_7$ | — |
| $B_5$ | $B_5$ |
| ... etc. | ... etc. |

I-pictures are useful resynchronisation points in the coded bit stream: because it is coded without prediction, an I-picture may be decoded independently of any other coded pictures. This supports random access by a decoder (a decoder may start decoding the bit stream at any I-picture position) and error resilience (discussed in Chapter 11). However, an I-picture has poor compression efficiency because no temporal prediction is used. P-pictures provide better compression efficiency due to motion-compensated prediction and can be used as prediction references. B-pictures have the highest compression efficiency of each of the three picture types.

The MPEG-1 standard does not actually define the design of an encoder: instead, the standard describes the coded syntax and a hypothetical 'reference' decoder. In practice, the syntax and functionality described by the standard mean that a compliant encoder has to contain certain functions. The basic CODEC is similar to Figure 3.18. A 'front end' carries out motion estimation and compensation based on one reference frame (P-pictures) or two reference frames (B-pictures). The motion-compensated residual (or the original picture data in the case of an I-picture) is encoded using DCT, quantisation, run-level coding and variable-length coding. In an I- or P-picture, quantised transform coefficients are rescaled and transformed with the inverse DCT to produce a stored reference frame for further predicted P- or B-pictures. In the decoder, the coded data is entropy decoded, rescaled, inverse transformed and motion compensated. The most complex part of the CODEC is often the motion estimator because bidirectional motion estimation is computationally intensive. Motion estimation is only required in the encoder and this is another example of asymmetry between the encoder and decoder.

## MPEG-1 syntax

The syntax of an MPEG-1 coded video sequence forms a hierarchy as shown in Figure 4.10. The levels or *layers* of the hierarchy are as follows.

**Sequence layer**   This may correspond to a complete encoded video programme. The sequence starts with a *sequence header* that describes certain key information about the coded sequence including picture resolution and frame rate. The sequence consists of a series of *groups of pictures* (GOPs), the next layer of the hierarchy.

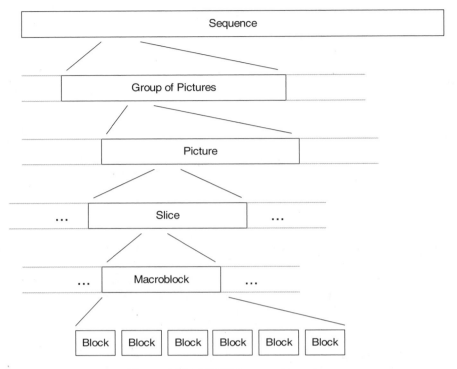

**Figure 4.10**  MPEG-1 synatx hierarchy

**GOP layer**  A GOP is one I-picture followed by a series of P- and B-pictures (e.g. Figure 4.8). In Figure 4.8, the GOP contains nine pictures (one I, two P and six B) but many other GOP structures are possible, for example:

(a) All GOPs contain just one I-picture, i.e. no motion compensated prediction is used: this is similar to Motion JPEG.

(b) GOPs contain only I- and P-pictures, i.e. no bidirectional prediction is used: compression efficiency is relatively poor but complexity is low (since B-pictures are more complex to generate).

(c) Large GOPs: the proportion of I-pictures in the coded stream is low and hence compression efficiency is high. However, there are few synchronisation points which may not be ideal for random access and for error resilience.

(d) Small GOPs: there is a high proportion of I-pictures and so compression efficiency is low, however there are frequent opportunities for resynchronisation.

An encoder need not keep a consistent GOP structure within a sequence. It may be useful to vary the structure occasionally, for example by starting a new GOP when a scene change or cut occurs in the video sequence.

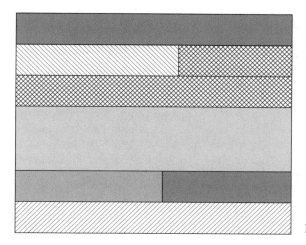

**Figure 4.11**   Example of MPEG-1 slices

**Picture layer**   A picture defines a single coded frame. The picture header describes the type of coded picture (I, P, B) and a temporal reference that defines when the picture should be displayed in relation to the other pictures in the sequence.

**Slice layer**   A picture is made up of a number of slices, each of which contains an integral number of macroblocks. In MPEG-1 there is no restriction on the size or arrangement of slices in a picture, except that slices should cover the picture in raster order. Figure 4.11 shows one possible arrangement: each shaded region in this figure is a single slice.

A slice starts with a slice header that defines its position. Each slice may be decoded independently of other slices within the picture and this helps the decoder to recover from transmission errors: if an error occurs within a slice, the decoder can always restart decoding from the next slice header.

**Macroblock layer**   A slice is made up of an integral number of macroblocks, each of which consists of six blocks (Figure 4.6). The macroblock header describes the type of macroblock, motion vector(s) and defines which $8 \times 8$ blocks actually contain coded transform data. The picture type (I, P or B) defines the 'default' prediction mode for each macroblock, but individual macroblocks within P- or B-pictures may be intra-coded if required (i.e. coded without any motion-compensated prediction). This can be useful if no good match can be found within the search area in the reference frames since it may be more efficient to code the macroblock without any prediction.

**Block layer**   A block contains variable-length code(s) that represent the quantised transform coefficients in an $8 \times 8$ block. Each DC coefficient (DCT coefficient [0, 0]) is coded differentially from the DC coefficient of the previous coded block, to exploit the fact that neighbouring blocks tend to have very similar DC (average) values. AC coefficients (all other coefficients) are coded as a (run, level) pair, where 'run' indicates the number of preceding zero coefficients and 'level' the value of a non-zero coefficient.

## 4.4.2   MPEG-2

The next important entertainment application for coded video (after CD-ROM storage) was digital television. In order to provide an improved alternative to analogue television, several key features were required of the video coding algorithm. It had to efficiently support larger frame sizes (typically $720 \times 576$ or $720 \times 480$ pixels for ITU-R 601 resolution) and coding of interlaced video. MPEG-1 was primarily designed to support progressive video, where each frame is scanned as a single unit in raster order. At television-quality resolutions, interlaced video (where a frame is made up of two interlaced 'fields' as described in Chapter 2) gives a smoother video image. Because the two fields are captured at separate time intervals (typically 1/50 or 1/60 of a second apart), better performance may be achieved by coding the fields separately.

MPEG-2 consists of three main sections: Video (described below), Audio[18] (based on MPEG-1 audio coding) and Systems[19] (defining, in more detail than MPEG-1 Systems, multiplexing and transmission of the coded audio/visual stream). MPEG-2 Video is (almost) a superset of MPEG-1 Video, i.e. most MPEG-1 video sequences should be decodeable by an MPEG-2 decoder. The main enhancements added by the MPEG-2 standard are as follows:

### Efficient coding of television-quality video

The most important application of MPEG-2 is broadcast digital television. The 'core' functions of MPEG-2 (described as 'main profile/main level') are optimised for efficient coding of television resolutions at a bit rate of around 3–5 Mbps.

### Support for coding of interlaced video

MPEG-2 video has several features that support flexible coding of interlaced video. The two fields that make up a complete interlaced frame can be encoded as separate pictures (*field pictures*), each of which is coded as an I-, P- or B-picture. P- and B- field pictures may be predicted from a field in another frame or from the other field in the current frame.

Alternatively, the two fields may be handled as a single picture (a *frame picture*) with the luminance samples in each macroblock of a frame picture arranged in one of two ways. *Frame DCT coding* is similar to the MPEG-1 structure, where each of the four luminance blocks contains alternate lines from both fields. With *field DCT coding*, the top two luminance blocks contain only samples from the top field, and the bottom two luminance blocks contain samples from the bottom field. Figure 4.12 illustrates the two coding structures.

In a field picture, the upper and lower $16 \times 8$ sample regions of a macroblock may be motion-compensated independently: hence each of the two regions has its own vector (or two vectors in the case of a B-picture). This adds an overhead to the macroblock because of the extra vector(s) that must be transmitted. However, this *16 × 8 motion compensation* mode can improve performance because a field picture has half the vertical resolution of a frame picture and so there are more likely to be significant differences in motion between the top and bottom halves of each macroblock.

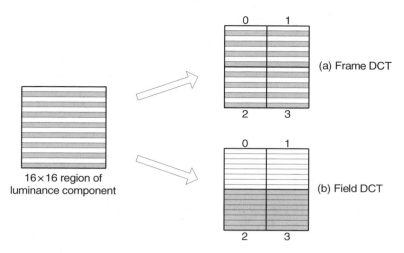

**Figure 4.12**   (a) Frame and (b) field DCT coding


In *dual-prime motion compensation* mode, the current field (within a field or frame picture) is predicted from the two fields of the reference frame using a single vector together with a transmitted correction factor. The correction factor modifies the motion vector to compensate for the small displacement between the two fields in the reference frame.


## *Scalability*

The progressive modes of JPEG described earlier are forms of *scalable coding*. A scalable coded bit stream consists of a number of layers, a *base layer* and one or more *enhancement layers*. The base layer can be decoded to provide a recognisable video sequence that has a limited visual quality, and a higher-quality sequence may be produced by decoding the base layer plus enhancement layer(s), with each extra enhancement layer improving the quality of the decoded sequence. MPEG-2 video supports four scalable modes.

**Spatial scalability**   This is analogous to hierarchical encoding in the JPEG standard. The base layer is coded at a low spatial resolution and each enhancement layer, when added to the base layer, gives a progressively higher spatial resolution.

**Temporal scalability**   The base layer is encoded at a low temporal resolution (frame rate) and the enhancement layer (s) are coded to provide higher frame rate(s) (Figure 4.13). One application of this mode is stereoscopic video coding: the base layer provides a monoscopic 'view' and an enhancement layer provides a stereoscopic offset 'view'. By combining the two layers, a full stereoscopic image may be decoded.

**SNR scalability**   In a similar way to the successive approximation mode of JPEG, the base layer is encoded at a 'coarse' visual quality (with high compression). Each enhancement layer, when added to the base layer, improves the video quality.

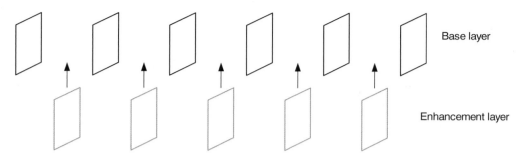

**Figure 4.13**  Temporal scalability

**Data partitioning**  The coded sequence is partitioned into two layers. The base layer contains the most 'critical' components of the coded sequence such as header information, motion vectors and (optionally) low-frequency transform coefficients. The enhancement layer contains all remaining coded data (usually less critical to successful decoding).

These scalable modes may be used in a number of ways. A decoder may decode the current programme at standard ITU-R 601 resolution (720 × 576 pixels, 25 or 30 frames per second) by decoding just the base layer, whereas a 'high definition' decoder may decode one or more enhancement layer (s) to increase the temporal and/or spatial resolution. The multiple layers can support simultaneous decoding by 'basic' and 'advanced' decoders. Transmission of the base and enhancement layers is usually more efficient than encoding and sending separate bit streams at the lower and higher resolutions.

   The base layer is the most 'important' to provide a visually acceptable decoded picture. Transmission errors in the base layer can have a catastrophic effect on picture quality, whereas errors in enhancement layer (s) are likely to have a relatively minor impact on quality. By protecting the base layer (for example using a separate transmission channel with a low error rate or by adding error correction coding), high visual quality can be maintained even when transmission errors occur (see Chapter 11).

*Profiles and levels*

Most applications require only a limited subset of the wide range of functions supported by MPEG-2. In order to encourage interoperability for certain 'key' applications (such as digital TV), the standard includes a set of recommended *profiles* and *levels* that each define a certain subset of the MPEG-2 functionalities. Each profile defines a set of *capabilities* and the important ones are as follows:

- *Simple*: 4 : 2 : 0 sampling, only I- and P-pictures are allowed. Complexity is kept low at the expense of poor compression performance.

- *Main*: This includes all of the core MPEG-2 capabilities including B-pictures and support for interlaced video. 4 : 2 : 0 sampling is used.

- *4 : 2 : 2*: As the name suggests, 4 : 2 : 2 subsampling is used, i.e. the Cr and Cb components have full vertical resolution and half horizontal resolution. Each macroblock contains eight blocks: four luminance, two Cr and two Cb.

- *SNR*: As 'main' profile, except that an enhancement layer is added to provide higher visual quality.

- *Spatial*: As 'SNR' profile, except that spatial scalability may also be used to provide higher-quality enhancement layers.

- *High*: As 'Spatial' profile, with the addition of support for $4:2:2$ sampling.

Each level defines *spatial and temporal resolutions*:

- *Low*: Up to $352 \times 288$ frame resolution and up to 30 frames per second.

- *Main*: Up to $720 \times 576$ frame resolution and up to 30 frames per second.

- *High–1440*: Up to $1440 \times 1152$ frame resolution and up to 60 frames per second.

- *High*: Up to $1920 \times 1152$ frame resolution and up to 60 frames per second.

The MPEG-2 standard defines certain recommended combinations of profiles and levels. *Main profile / low level* (using only frame encoding) is essentially MPEG-1. *Main profile / main level* is suitable for broadcast digital television and this is the most widely used profile / level combination. *Main profile / high level* is suitable for high-definition television (HDTV). (Originally, the MPEG working group intended to release a further standard, MPEG-3, to support coding for HDTV applications. However, once it became clear that the MPEG-2 syntax could deal with this application adequately, work on this standard was dropped and so there is no MPEG-3 standard.)

In addition to the main features described above, there are some further changes from the MPEG-1 standard. Slices in an MPEG-2 picture are constrained such that they may not overlap from one row of macroblocks to the next (unlike MPEG-1 where a slice may occupy multiple rows of macroblocks). D-pictures in MPEG-1 were felt to be of limited benefit and are not supported in MPEG-2.

### 4.4.3   MPEG-4

The MPEG-1 and MPEG-2 standards deal with complete video frames, each coded as a single unit. The MPEG-4 standard[6] was developed with the aim of extending the capabilities of the earlier standards in a number of ways.

**Support for low bit-rate applications**   MPEG-1 and MPEG-2 are reasonably efficient for coded bit rates above around 1 Mbps. However, many emerging applications (particularly Internet-based applications) require a much lower transmission bit rate and MPEG-1 and 2 do not support efficient compression at low bit rates (tens of kbps or less).

**Support for object-based coding**   Perhaps the most fundamental shift in the MPEG-4 standard has been towards *object-based* or *content-based* coding, where a video scene can be handled as a set of foreground and background *objects* rather than just as a series of rectangular frames. This type of coding opens up a wide range of possibilities, such as independent coding of different objects in a scene, reuse of scene components, compositing

**Figure 4.14**  Video scene
showing multiple video objects

(where objects from a number of sources are combined into a scene) and a high degree of
interactivity. The basic concept used in MPEG-4 Visual is that of the *video object* (VO). A
video scene (VS) (a sequence of video frames) is made up of a number of VOs. For example,
the VS shown in Figure 4.14 consists of a background VO and two foreground VOs. MPEG-4
provides tools that enable each VO to be coded independently, opening up a range of new
possibilities. The equivalent of a 'frame' in VO terms, i.e. a 'snapshot' of a VO at a single
instant in time, is a *video object plane* (VOP). The entire scene may be coded as a single,
rectangular VOP and this is equivalent to a picture in MPEG-1 and MPEG-2 terms.

**Toolkit-based coding**   MPEG-1 has a very limited degree of flexibility; MPEG-2 introduced the concept of a 'toolkit' of profiles and levels that could be combined in different
ways for various applications. MPEG-4 extends this towards a highly flexible set of coding
tools that enable a range of applications as well as a standardised framework that allows new
tools to be added to the 'toolkit'.

The MPEG-4 standard is organised so that new coding tools and functionalities may be
added incrementally as new versions of the standard are developed, and so the list of tools
continues to grow. However, the main tools for coding of video images can be summarised
as follows.

### *MPEG-4 Visual: very low bit-rate video core*

The video coding algorithms that form the 'very low bit-rate video (VLBV) core' of MPEG-
4 Visual are almost identical to the baseline H.263 video coding standard (Chapter 5). If the
*short header* mode is selected, frame coding is completely identical to baseline H.263. A
video sequence is coded as a series of rectangular frames (i.e. a single VOP occupying the
whole frame).

**Input format**   Video data is expected to be pre-processed and converted to one of the
picture sizes listed in Table 4.2, at a frame rate of up to 30 frames per second and in $4:2:0$
$Y:Cr:Cb$ format (i.e. the chrominance components have half the horizontal and vertical
resolution of the luminance component).

**Picture types**   Each frame is coded as an I- or P-frame. An I-frame contains only intra-
coded macroblocks, whereas a P-frame can contain either intra- or inter-coded macroblocks.

**Table 4.2** MPEG4 VLBV/H.263 picture sizes

| Format | Picture size (luminance) |
|--------|--------------------------|
| SubQCIF | $128 \times 96$ |
| QCIF | $176 \times 144$ |
| CIF | $352 \times 288$ |
| 4CIF | $704 \times 576$ |
| 16CIF | $1408 \times 1152$ |

**Motion estimation and compensation**  This is carried out on $16 \times 16$ macroblocks or (optionally) on $8 \times 8$ macroblocks. Motion vectors can have half-pixel resolution.

**Transform coding**  The motion-compensated residual is coded with DCT, quantisation, zigzag scanning and run–level coding.

**Variable-length coding**  The run–level coded transform coefficients, together with header information and motion vectors, are coded using variable-length codes. Each non-zero transform coefficient is coded as a combination of *run, level, last* (where 'last' is a flag to indicate whether this is the last non-zero coefficient in the block) (see Chapter 8).

*Syntax*

The syntax of an MPEG-4 (VLBV) coded bit stream is illustrated in Figure 4.15.

**Picture layer**  The highest layer of the syntax contains a complete coded picture. The picture header indicates the picture resolution, the type of coded picture (inter or intra) and includes a temporal reference field. This indicates the correct display time for the decoder (relative to other coded pictures) and can help to ensure that a picture is not displayed too early or too late.

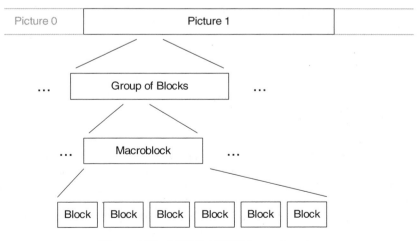

**Figure 4.15**  MPEG-4/H.263 layered syntax

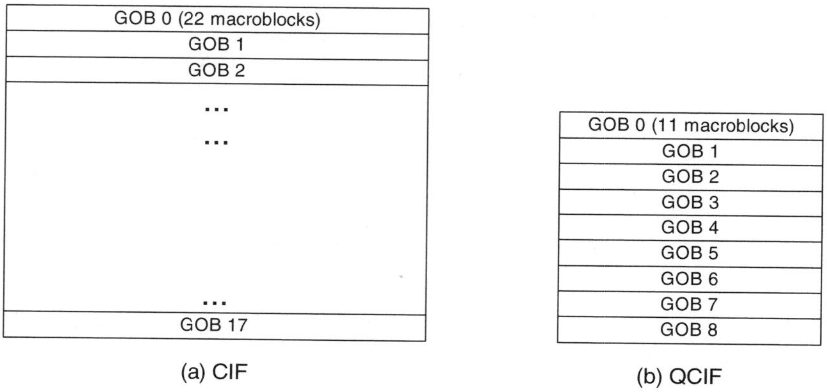

| GOB 0 (22 macroblocks) |
| GOB 1 |
| GOB 2 |
| ... |
| ... |
| |
| |
| ... |
| GOB 17 |

| GOB 0 (11 macroblocks) |
| GOB 1 |
| GOB 2 |
| GOB 3 |
| GOB 4 |
| GOB 5 |
| GOB 6 |
| GOB 7 |
| GOB 8 |

(a) CIF                                              (b) QCIF

**Figure 4.16**    GOBs: (a) CIF and (b) QCIF pictures

**Group of blocks layer**    A group of blocks (GOB) consists of one complete row of macro-blocks in SQCIF, QCIF and CIF pictures (two rows in a 4CIF picture and four rows in a 16CIF picture). GOBs are similar to slices in MPEG-1 and MPEG-2 in that, if an optional GOB header is inserted in the bit stream, the decoder can resynchronise to the start of the next GOB if an error occurs. However, the size and layout of each GOB are fixed by the standard (unlike slices). The arrangement of GOBs in a QCIF and CIF picture is shown in Figure 4.16.

**Macroblock layer**    A macroblock consists of four luminance blocks and two chrominance blocks. The macroblock header includes information about the type of macroblock, 'coded block pattern' (indicating which of the six blocks actually contain transform coefficients) and coded horizontal and vertical motion vectors (for inter-coded macroblocks).

**Block layer**    A block consists of run–level coded coefficients corresponding to an $8 \times 8$ block of samples.

The core CODEC (based on H.263) was designed for efficient coding at low bit rates. The use of $8 \times 8$ block motion compensation and the design of the variable-length coding tables make the VLBV MPEG-4 CODEC more efficient than MPEG-1 or MPEG-2 (see Chapter 5 for a comparison of coding efficiency).

## Other visual coding tools

The features that make MPEG-4 (Visual) unique among the coding standards are the range of further coding tools available to the designer.

**Shape coding**    Shape coding is required to specify the boundaries of each non-rectangular VOP in a scene. Shape information may be *binary* (i.e. identifying the pixels that are internal to the VOP, described as 'opaque', or external to the VOP, described as 'transparent') or *grey scale* (where each pixel position within a VOP is allocated an 8-bit 'grey scale' number that identifies the transparency of the pixel). Grey scale information is more complex and requires more bits to code: however, it introduces the possibility of overlapping, semi-transparent VOPs (similar to the concept of 'alpha planes' in computer graphics). Binary information is simpler to code because each pixel has only two possible states, opaque or transparent. Figure 4.17

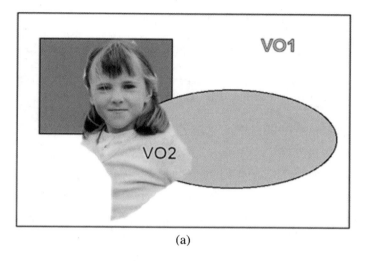

(a)

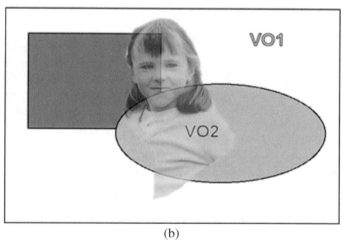

(b)

**Figure 4.17** (a) Opaque and (b) semi-transparent VOPs

illustrates the concept of opaque and semi-transparent VOPs: in image (a), VOP2 (foreground) is opaque and completely obscures VOP1 (background), whereas in image (b) VOP2 is partly transparent.

Binary shape information is coded in $16 \times 16$ blocks (binary alpha blocks, BABs). There are three possibilities for each block:

1. All pixels are transparent, i.e. the block is 'outside' the VOP. No shape (or texture) information is coded.

2. All pixels are opaque, i.e. the block is fully 'inside' the VOP. No shape information is coded: the pixel values of the block ('texture') are coded as described in the next section.

3. Some pixels are opaque and some are transparent, i.e. the block crosses a boundary of the VOP. The binary shape values of each pixel (1 or 0) are coded using a form of DPCM and the texture information of the opaque pixels is coded as described below.

Grey scale shape information produces values in the range 0 (transparent) to 255 (opaque) that are compressed using block-based DCT and motion compensation.

**Motion compensation**    Similar options exist to the I-, P- and B-pictures in MPEG-1 and MPEG-2:

1. I-VOP: VOP is encoded without any motion compensation.

2. P-VOP: VOP is predicted using motion-compensated prediction from a past I- or P-VOP.

3. B-VOP: VOP is predicted using motion-compensated prediction from a past and a future I- or P-picture (with forward, backward or bidirectional prediction).

Figure 4.18 shows mode (3), prediction of a B-VOP from a previous I-VOP and future P-VOP. For macroblocks (or $8 \times 8$ blocks) that are fully contained within the current and reference VOPs, block-based motion compensation is used in a similar way to MPEG-1 and MPEG-2. The motion compensation process is modified for blocks or macroblocks along the boundary of the VOP. In the reference VOP, pixels in the $16 \times 16$ (or $8 \times 8$) search area are padded based on the pixels along the edge of the VOP. The macroblock (or block) in the current VOP is matched with this search area using block matching: however, the difference value (mean absolute error or sum of absolute errors) is only computed for those pixel positions that lie within the VOP.

**Texture coding**    Pixels (or motion-compensated residual values) within a VOP are coded as 'texture'. The basic tools are similar to MPEG-1 and MPEG-2: transform using the DCT, quantisation of the DCT coefficients followed by reordering and variable-length coding. To further improve compression efficiency, quantised DCT coefficients may be *predicted* from previously transmitted blocks (similar to the differential prediction of DC coefficients used in JPEG, MPEG-1 and MPEG-2).

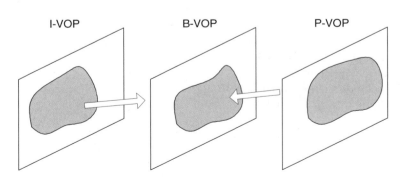

**Figure 4.18**    B-VOP motion-compensated prediction

A macroblock that covers a boundary of the VOP will contain both opaque and transparent pixels. In order to apply a regular $8 \times 8$ DCT, it is necessary to use 'padding' to fill up the transparent pixel positions. In an inter-coded VOP, where the texture information is motion-compensated residual data, the transparent positions are simply filled with zeros. In an intra-coded VOP, where the texture is 'original' pixel data, the transparent positions are filled by extrapolating the pixel values along the boundary of the VOP.

**Error resilience**   MPEG-4 incorporates a number of mechanisms that can provide improved performance in the presence of transmission errors (such as bit errors or lost packets). The main tools are:

1. Synchronisation markers: similar to MPEG-1 and MPEG-2 slice start codes, except that these may optionally be positioned so that each resynchronisation interval contains an approximately equal number of encoded bits (rather than a constant number of macroblocks). This means that errors are likely to be evenly distributed among the resynchronisation intervals. Each resynchronisation interval may be transmitted in a separate *video packet*.

2. Data partitioning: similar to the data partitioning mode of MPEG-2.

3. Header extension: redundant copies of header information are inserted at intervals in the bit stream so that if an important header (e.g. a picture header) is lost due to an error, the redundant header may be used to partially recover the coded scene.

4. Reversible VLCs: these variable length codes limit the propagation ('spread') of an errored region in a decoded frame or VOP and are described further in Chapter 8.

**Scalability**   MPEG-4 supports spatial and temporal scalability. Spatial scalability applies to rectangular VOPs in a similar way to MPEG-2: the base layer gives a low spatial resolution and an enhancement layer may be decoded together with the base layer to give a higher resolution. Temporal scalability is extended beyond the MPEG-2 approach in that it may be applied to individual VOPs. For example, a background VOP may be encoded without scalability, whilst a foreground VOP may be encoded with several layers of temporal scalability. This introduces the possibility of decoding a foreground object at a higher frame rate and more static, background objects at a lower frame rate.

**Sprite coding**   A 'sprite' is a VOP that is present for the entire duration of a video sequence (VS). A sprite may be encoded and transmitted once at the start of the sequence, giving a potentially large benefit in compression performance. A good example is a background sprite: the background image to a scene is encoded as a sprite at the start of the VS. For the remainder of the VS, only the foreground VOPs need to be coded and transmitted since the decoder can 'render' the background from the original sprite. If there is camera movement (e.g. panning), then a sprite that is larger than the visible scene is required (Figure 4.19). In order to compensate for more complex camera movements (e.g. zoom or rotation), it may be necessary for the decoder to 'warp' the sprite. A sprite is encoded as an I-VOP as described earlier.

**Static texture**   An alternative set of tools to the DCT may be used to code 'static' texture, i.e. texture data that does not change rapidly. The main application for this is to code texture

**Figure 4.19**   Example of background sprite and foreground VOPs

that is mapped onto a 2-D or 3-D surface (described below). Static image texture is coded efficiently using a wavelet transform. The transform coefficients are quantised and coded with a zero-tree algorithm followed by arithmetic coding. Wavelet coding is described further in Chapter 7 and arithmetic coding in Chapter 8.

**Mesh and 3-D model coding**   MPEG-4 supports more advanced object-based coding techniques including:

- 2-D mesh coding, where an object is coded as a mesh of triangular patches in a 2-D plane. Static texture (coded as described above) can be mapped onto the mesh. A moving object can be represented by deforming the mesh and warping the texture as the mesh moves.

- 3-D mesh coding, where an object is described as a mesh in 3-D space. This is more complex than a 2-D mesh representation but gives a higher degree of flexibility in terms of representing objects within a scene.

- Face and body model coding, where a human face or body is rendered at the decoder according to a face or body model. The model is controlled (moved) by changing 'animation parameters'. In this way a 'head-and-shoulders' video scene may be coded by sending only the animation parameters required to 'move' the model at the decoder. Static texture is mapped onto the model surface.

These three tools offer the potential for fundamental improvements in video coding performance and flexibility: however, their application is currently limited because of the high processing resources required to analyse and render even a very simple scene.

*MPEG-4 visual profiles and levels*

In common with MPEG-2, a number of recommended 'profiles' (sets of MPEG-4 tools) and 'levels' (constraints on bit stream parameters such as frame size and rate) are defined in the

MPEG-4 standard. Each profile is defined in terms of one or more 'object types', where an object type is a subset of the MPEG-4 tools. Table 4.3 lists the main MPEG-4 object types that make up the profiles. The 'Simple' object type contains tools for coding of basic I- and P-rectangular VOPs (complete frames) together with error resilience tools and the 'short header' option (for compatibility with H.263). The 'Core' type adds B-VOPs and basic shape coding (using a binary shape mask only). The main profile adds grey scale shape coding and sprite coding.

MPEG-4 (Visual) is gaining popularity in a number of application areas such as Internet-based video. However, to date the majority of applications use only the simple object type and there has been limited take-up of the content-based features of the standard. This is partly because of technical complexities (for example, it is difficult to accurately segment a video scene into foreground and background objects, e.g. Figure 4.14, using an automatic algorithm) and partly because useful applications for content-based video coding and manipulation have yet to emerge. At the time of writing, the great majority of video coding applications continue to work with complete rectangular frames. However, researchers continue to improve algorithms for segmenting and manipulating video objects.[20–25] The content-based tools have a number of interesting possibilities: for example, they make it

**Table 4.3**　MPEG-4 video object types

| Visual tools | Video object types | | | | | | | |
| --- | --- | --- | --- | --- | --- | --- | --- | --- |
| | Simple | Core | Main | Simple scalable | Animated 2-D mesh | Basic animated texture | Still scalable texture | Simple face |
| Basic (I-VOP, P-VOP, coefficient prediction, 16 × 16 and 8 × 8 motion vectors) | ✓ | ✓ | ✓ | ✓ | ✓ | | | |
| Error resilience | ✓ | ✓ | ✓ | ✓ | ✓ | | | |
| Short header | ✓ | ✓ | ✓ | | ✓ | | | |
| B-VOP | | ✓ | ✓ | ✓ | ✓ | | | |
| P-VOP with overlapped block matching | | | | | | | | |
| Alternative quantisation | | ✓ | ✓ | | ✓ | | | |
| P-VOP based temporal scalability | | ✓ | ✓ | | ✓ | | | |
| Binary shape | | ✓ | ✓ | | ✓ | ✓ | | |
| Grey shape | | | ✓ | | | | | |
| Interlaced video coding | | | ✓ | | | | | |
| Sprite | | | ✓ | | | | | |
| Rectangular temporal scalability | | | | ✓ | | | | |
| Rectangular spatial scalability | | | | ✓ | | | | |
| Scalable still texture | | | | | ✓ | ✓ | ✓ | |
| 2-D mesh | | | | | ✓ | ✓ | | |
| Facial animation parameters | | | | | | | | ✓ |

possible to develop 'hybrid' applications with a mixture of 'real' video objects (possibly from a number of different sources) and computer-generated graphics. So-called synthetic natural hybrid coding has the potential to enable a new generation of video applications.

## 4.5  SUMMARY

The ISO has issued a number of image and video coding standards that have heavily influenced the development of the technology and market for video coding applications. The original JPEG still image compression standard is now a ubiquitous method for storing and transmitting still images and has gained some popularity as a simple and robust algorithm for video compression. The improved subjective and objective performance of its successor, JPEG-2000, may lead to the gradual replacement of the original JPEG algorithm.

The first MPEG standard, MPEG-1, was never a market success in its target application (video CDs) but is widely used for PC and internet video applications and formed the basis for the MPEG-2 standard. MPEG-2 has enabled a worldwide shift towards digital television and is probably the most successful of the video coding standards in terms of market penetration. The MPEG-4 standard offers a plethora of video coding tools which may in time enable many new applications: however, at the present time the most popular element of MPEG-4 (Visual) is the 'core' low bit rate CODEC that is based on the ITU-T H.263 standard. In the next chapter we will examine the H.26x series of coding standards, H.261, H.263 and the emerging H.26L.

## REFERENCES

1. http://www.itu.int/ [International Telecommunication Union].
2. http://www.iso.ch/ [International Standards Organisation].
3. ISO/IEC 10918-1/ITU-T Recommendation T.81, 'Digital compression and coding of continuous-tone still images', 1992 [JPEG].
4. ISO/IEC 11172-2, 'Information technology–coding of moving pictures and associated audio for digital storage media at up to about 1.5 Mbit/s–part 2: Video', 1993 [MPEG1 Video].
5. ISO/IEC 13818-2, 'Information technology: generic coding of moving pictures and associated audio information: Video', 1995 [MPEG2 Video].
6. ISO/IEC 14996-2, 'Information technology–coding of audio-visual objects–part 2: Visual', 1998 [MPEG-4 Visual].
7. ISO/IEC FCD 15444-1, 'JPEG2000 Final Committee Draft v1.0', March 2000.
8. ISO/IEC JTC1/SC29/WG11 N4031, 'Overview of the MPEG-7 Standard', Singapore, March 2001.
9. ISO/IEC JTC1/SC29/WG11 N4318, 'MPEG-21 Overview', Sydney, July 2001.
10. http://standards.pictel.com/ftp/video-site/ [VCEG working documents].
11. http://www.cselt.it/mpeg/ [MPEG committee official site].
12. http://www.jpeg.org/ [JPEG resources].
13. http://www.mpeg.org/ [MPEG resources].
14. ITU-T Q6/SG16 Draft Document, 'Appendix III for ITU-T Rec H.263', Porto Seguro, May 2001.
15. A. N. Skodras, C. A. Christopoulos and T. Ebrahimi, 'JPEG2000: The upcoming still image compression standard', *Proc. 11th Portuguese Conference on Pattern Recognition*, Porto, 2000.
16. ISO/IEC 11172-1, 'Information technology–coding of moving pictures and associated audio for digital storage media at up to about 1.5 Mbit/s–part 1: Systems', 1993 [MPEG1 Systems].

17. ISO/IEC 11172-2, Information technology–coding of moving pictures and associated audio for digital storage mediat at up to about 1.5 Mbit/s–part 3: Audio', 1993 [MPEG1 Audio].

18. ISO/IEC 13818-3, 'Information technology: generic coding of moving pictures and associated audio information: Audio', 1995 [MPEG2 Audio].

19. ISO/IEC 13818-1, 'Information technology: generic coding of moving pictures and associated audio information Systems', 1995 [MPEG2 Systems].

20. P. Salembier and F. Marqués, 'Region-based representations of image and video: segmentation tools for multimedia services', *IEEE Trans. CSVT* **9**(8), December 1999.

21. L. Garrido, A. Oliveras and P. Salembier, 'Motion analysis of image sequences using connected operators', *Proc. VCIP97*, San Jose, February 1997, *SPIE* **3024**.

22. K. Illgner and F. Müller, 'Image segmentation using motion estimation', in *Time-varying Image Processing and Image Recognition*, Elsevier Science, 1997.

23. R. Castagno and T. Ebrahimi, 'Video Segmentation based on multiple features for interactive multimedia applications', *IEEE Trans. CSVT* **8**(5), September, 1998.

24. E. Steinbach, P. Eisert and B. Girod, 'Motion-based analysis and segmentation of image sequences using 3-D scene models', *Signal Processing*, **66**(2), April 1998.

25. M. Chang, M. Teklap and M. Ibrahim Sezan, 'Simultaneous motion estimation and segmentation', *IEEE Trans. Im. Proc.*, **6**(9), 1997.

# 5

# Video Coding Standards: H.261, H.263 and H.26L

## 5.1 INTRODUCTION

The ISO MPEG video coding standards are aimed at storage and distribution of video for entertainment and have tried to meet the needs of providers and consumers in the 'media industries'. The ITU has (historically) been more concerned about the telecommunications industry, and its video coding standards (H.261, H.263, H.26L) have consequently been targeted at real-time, point-to-point or multi-point communications.

The first ITU-T video coding standard to have a significant impact, H.261, was developed during the late 1980s/early 1990s with a particular application and transmission channel in mind. The application was video conferencing (two-way communications via a video 'link') and the channel was N-ISDN. ISDN provides a constant bit rate of $p \times 64$ kbps, where $p$ is an integer in the range 1–30: it was felt at the time that ISDN would be the medium of choice for video communications because of its guaranteed bandwidth and low delay. Modem channels over the analogue POTS/PSTN (at speeds of less than 9600 bps at the time) were considered to be too slow for visual communications and packet-based transmission was not considered to be reliable enough.

H.261 was quite successful and continues to be used in many legacy video conferencing applications. Improvements in processor performance, video coding techniques and the emergence of analogue Modems and Internet Protocol (IP) networks as viable channels led to the development of its successor, H.263, in the mid-1990s. By making a number of improvements to H.261, H.263 provided significantly better compression performance as well as greater flexibility. The original H.263 standard (Version 1) had four optional modes which could be switched on to improve performance (at the expense of greater complexity). These modes were considered to be useful and Version 2 ('H.263+') added 12 further optional modes. The latest (and probably the last) version (v3) will contain a total of 19 modes, each offering improved coding performance, error resilience and/or flexibility.

Version 3 of H.263 has become a rather unwieldy standard because of the large number of options and the need to continue to support the basic ('baseline') CODEC functions. The latest initiative of the ITU-T experts group VCEG is the H.26L standard (where 'L' stands for 'long term'). This is a new standard that makes use of some of the best features of H.263 and aims to improve compression performance by around 50% at lower bit rates. Early indications are that H.26L will outperform H.263+ (but possibly not by 50%).

## 5.2  H.261[1]

Typical operating bit rates for H.261 applications are between 64 and 384 kbps. At the time of development, packet-based transmission over the Internet was not expected to be a significant requirement, and the limited video compression performance achievable at the time was not considered to be sufficient to support bit rates below 64 kbps.

A typical H.261 CODEC is very similar to the 'generic' motion-compensated DCT-based CODEC described in Chapter 3. Video data is processed in $4:2:0$ $Y:Cr:Cb$ format. The basic unit is the 'macroblock', containing four luminance blocks and two chrominance blocks (each $8 \times 8$ samples) (see Figure 4.6). At the input to the encoder, $16 \times 16$ macroblocks may be (optionally) motion compensated using integer motion vectors. The motion-compensated residual data is coded with an $8 \times 8$ DCT followed by quantisation and zigzag reordering. The reordered transform coefficients are run–level coded and compressed with an entropy encoder (see Chapter 8).

Motion compensation performance is improved by use of an optional *loop filter*, a 2-D spatial filter that operates on each $8 \times 8$ block in a macroblock prior to motion compensation (if the filter is switched on). The filter has the effect of 'smoothing' the reference picture which can help to provide a better prediction reference. Chapter 9 discusses loop filters in more detail (see for example Figures 9.11 and 9.12).

In addition, a forward error correcting code is defined in the standard that should be inserted into the transmitted bit stream. In practice, this code is often omitted from practical implementations of H.261: the error rate of an ISDN channel is low enough that error correction is not normally required, and the code specified in the standard is not suitable for other channels (such as a noisy wireless channel or packet-based transmission).

Each macroblock may be coded in 'intra' mode (no motion-compensated prediction) or 'inter' mode (with motion-compensated prediction). Only two frame sizes are supported, CIF ($352 \times 288$ pixels) and QCIF ($176 \times 144$ pixels).

H.261 was developed at a time when hardware and software processing performance was limited and therefore has the advantage of low complexity. However, its disadvantages include poor compression performance (with poor video quality at bit rates of under about 100 kbps) and lack of flexibility. It has been superseded by H.263, which has higher compression efficiency and greater flexibility, but is still widely used in installed video conferencing systems.

## 5.3  H.263[2]

In developing the H.263 standard, VCEG aimed to improve upon H.261 in a number of areas. By taking advantage of developments in video coding algorithms and improvements in processing performance, it provides better compression. H.263 provides greater flexibility than H.261: for example, a wider range of frame sizes is supported (listed in Table 4.2). The first version of H.263 introduced four optional modes, each described in an annex to the standard, and further optional modes were introduced in Version 2 of the standard ('H.263+'). The target application of H.263 is low-bit-rate, low-delay two-way video communications. H.263 can support video communications at bit rates below 20 kbps (at a very limited visual quality) and is now widely used both in 'established' applications such as video telephony and video conferencing and an increasing number of new applications (such as Internet-based video).

### 5.3.1 Features

The baseline H.263 CODEC is functionally identical to the MPEG-4 'short header' CODEC described in Section 4.4.3. Input frames in $4:2:0$ format are motion compensated (with half-pixel resolution motion vectors), transformed with an $8 \times 8$ DCT, quantised, reordered and entropy coded. The main factors that contribute to the improved coding performance over H.261 are the use of half-pixel motion vectors (providing better motion compensation) and redesigned variable-length code (VLC) tables (described further in Chapter 8). Features such as I- and P-pictures, more frame sizes and optional coding modes give the designer greater flexibility to deal with different application requirements and transmission scenarios.

## 5.4 THE H.263 OPTIONAL MODES/H.263+

The original H.263 standard (Version 1) included four optional coding modes (Annexes D, E, F and G). Version 2 of the standard added 12 further modes (Annexes I to T) and a new release is scheduled with yet more coding modes (Annexes U, V and W). CODECs that implement some of the optional modes are sometimes described as 'H.263+' or 'H.263++' CODECs depending on which modes are implemented.

Each mode adds to or modifies the functionality of H.263, usually at the expense of increased complexity. An H.263-compliant CODEC must support the 'baseline' syntax described above: the use of optional modes may be negotiated between an encoder and a decoder prior to starting a video communications session. The optional modes have a number of potential benefits: some of the modes improve compression performance, others improve error resilience or provide tools that are useful for particular transmission environments such as packet-based transmission.

**Annex D, Unrestricted motion vectors** The optional mode described in Annex D of H.263 allows motion vectors to point outside the boundaries of the picture. This can provide a coding performance gain, particularly if objects are moving into or out of the picture. The pixels at the edges of the picture are extrapolated to form a 'border' outside the picture that vectors may point to (Figure 5.1). In addition, the motion vector range is extended so that

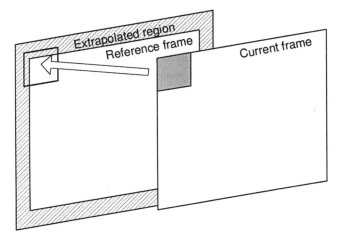

**Figure 5.1** Unrestricted motion vectors

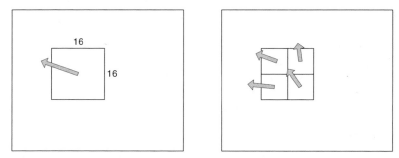

**Figure 5.2**    One or four motion vectors per macroblock

longer vectors are allowed. Finally, Annex D contains an optional alternative set of VLCs for encoding motion vector data. These VLCs are reversible, making it easier to recover from transmission errors (see Chapter 11).

**Annex E, Syntax-based arithmetic coding**    Arithmetic coding is used instead of variable-length coding. Each of the VLCs defined in the standard is replaced with a probability value that is used by an arithmetic coder (see Chapter 8).

**Annex F, Advanced prediction**    The efficiency of motion estimation and compensation is improved by allowing the use of four vectors per macroblock (a separate motion vector for each $8 \times 8$ luminance block, Figure 5.2). Overlapped block motion compensation (described in Chapter 6) is used to improve motion compensation and reduce 'blockiness' in the decoded image. Annex F requires the CODEC to support unrestricted motion vectors (Annex D).

**Annex G, PB-frames**    A PB-frame is a pair of frames coded as a combined unit. The first frame is coded as a 'B-picture' and the second as a P-picture. The P-picture is forward predicted from the previous I- or P-picture and the B-picture is bidirectionally predicted from the previous and current I- or P-pictures. Unlike MPEG-1 (where a B-picture is coded as a separate unit), each macroblock of the PB-frame contains data from both the P-picture and the B-picture (Figure 5.3). PB-frames can give an improvement in compression efficiency.

**Annex I, Advanced intra-coding**    This mode exploits the correlation between DCT coefficients in neighbouring intra-coded blocks in an image. The DC coefficient and the first row or column of AC coefficients may be predicted from the coefficients of neighbouring blocks (Figure 5.4). The zigzag scan, quantisation procedure and variable-length code tables are modified and the result is an improvement in compression efficiency for intra-coded macroblocks.

**Annex J, Deblocking filter**    The edges of each $8 \times 8$ block are 'smoothed' using a spatial filter (described in Chapter 9). This reduces 'blockiness' in the decoded picture and also improves motion compensation performance. When the deblocking filter is switched on, four

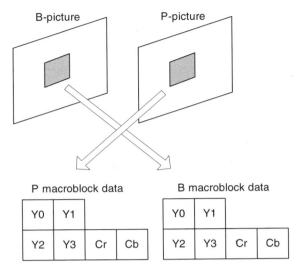

Figure 5.3 Macroblock in PB-frame

motion vectors per macroblock and unrestricted motion vectors are also enabled (Annexes D and F).

**Annex K, Slice structured mode** This mode provides support for resynchronisation intervals that are similar to MPEG-1 'slices'. A slice is a series of coded macroblocks

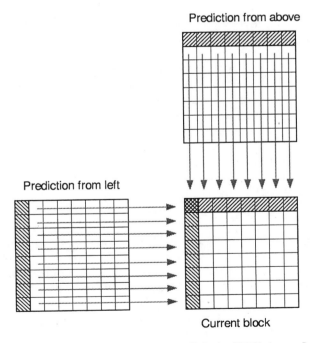

Figure 5.4 Prediction of intra-coefficients, H.263 Annex I

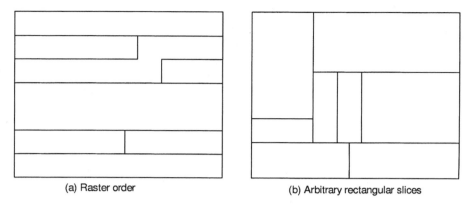

(a) Raster order                (b) Arbitrary rectangular slices

**Figure 5.5**    H.263 Annex K: slice options

starting with a slice header. Slices may contain macroblocks in raster order, or in any rectangular region of the picture (Figure 5.5). Slices may optionally be sent in an arbitrary order. Each slice may be decoded independently of any other slice in the picture and so slices can be useful for error resilience (see Chapter 11) since an error in one slice will not affect the decoding of any other slice.

**Annex L, Supplemental enhancement information**    This annex contains a number of supplementary codes that may be sent by an encoder to a decoder. These codes indicate display-related information about the video sequence, such as picture freeze and timing information.

**Annex M, Improved PB-frames**    As the name suggests, this is an improved version of the original PB-frames mode (Annex G). Annex M adds the options of forward or backward prediction for the B-frame part of each macroblock (as well as the bidirectional prediction defined in Annex G), resulting in improved compression efficiency.

**Annex N, Reference picture selection**    This mode enables an encoder to choose from a number of previously coded pictures for predicting the current picture. The use of this mode to limit error propagation in a noisy transmission environment is discussed in Chapter 11. At the start of each GOB or slice, the encoder may choose the preferred reference picture for prediction of macroblocks in that GOB or slice.

**Annex O, Scalability**    Temporal, spatial and SNR scalability are supported by this optional mode. In a similar way to the MPEG-2 optional scalability modes, spatial scalability increases frame resolution, SNR scalability increases picture quality and temporal scalability increases frame rate. In each case, a 'base layer' provides basic performance and the increased performance is obtained by decoding the base layer together with an "enhancement layer'. Temporal scalability is particularly useful because it supports B-pictures: these are similar to the 'true' B-pictures in the MPEG standards (where a B-picture is a separate coded unit) and are more flexible than the combined PB-frames described in Annexes G and M.

**Annex P, Reference picture resampling**   The prediction reference frame used by the encoder and decoder may be resampled prior to motion compensation. This has several possible applications. For example, an encoder can change the frame resolution 'on the fly' whilst continuing to use motion-compensated prediction. The prediction reference frame is resampled to match the new resolution and the current frame can then be predicted from the resampled reference. This mode may also be used to support *warping*, i.e. the reference picture is warped (deformed) prior to prediction, perhaps to compensate for nonlinear camera movements such as zoom or rotation.

**Annex Q, Reduced resolution update**   An encoder may choose to update selected macroblocks at a lower resolution than the normal spatial resolution of the frame. This may be useful, for example, to enable a CODEC to refresh moving parts of a frame at a low resolution using a small number of coded bits whilst keeping the static parts of the frame at the original higher resolution.

**Annex R, Independent segment decoding**   This annex extends the concept of the independently decodeable slices (Annex K) or GOBs. Segments of the picture (where a segment is one slice or an integral number of GOBs) may be decoded completely independently of any other segment. In the slice structured mode (Annex K), motion vectors can point to areas of the reference picture that are outside the current slice; with independent segment decoding, motion vectors and other predictions can only reference areas within the current segment in the reference picture (Figure 5.6). A segment can be decoded (over a series of frames) independently of the rest of the frame.

**Annex S, Alternative inter-VLC**   The encoder may use an alternative variable-length code table for transform coefficients in inter-coded blocks. The alternative VLCs (actually the same VLCs used for intra-coded blocks in Annex I) can provide better coding efficiency when there are a large number of high-valued quantised DCT coefficients (e.g. if the coded bit rate is high and/or there is a lot of variation in the video scene).

**Annex T, Modified quantisation**   This mode introduces some changes to the way the quantiser and rescaling operations are carried out. Annex T allows the encoder to change the

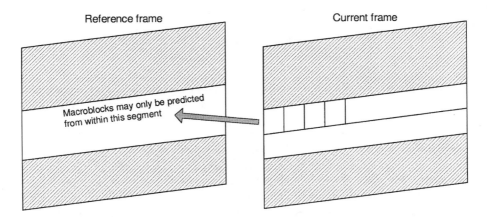

**Figure 5.6**   Independent segments

quantiser scale factor in a more flexible way during encoding, making it possible to control the encoder output bit rate more accurately.

**Annex U, Enhanced reference picture selection**   Annex U modifies the reference picture selection mode of Annex N to provide improved error resilience and coding efficiency. There are a number of changes, including a mechanism to reduce the memory requirements for storing previously coded pictures and the ability to select a reference picture for motion compensation on a macroblock-by-macroblock basis. This means that the 'best' match for each macroblock may be selected from any of a number of stored previous pictures (also known as *long-term memory prediction*).

**Annex V, Data partitioned slice**   Modified from Annex K, this mode improves the resilience of slice structured data to transmission errors. Within each slice, the macroblock data is rearranged so that all of the macroblock headers are transmitted first, followed by all of the motion vectors and finally by all of the transform coefficient data. An error occurring in header or motion vector data usually has a more serious effect on the decoded picture than an error in transform coefficient data: by rearranging the data in this way, an error occurring part-way through a slice should only affect the less-sensitive transform coefficient data.

**Annex W, Additional supplemental enhancement information**   Two extra enhancement information items are defined (in addition to those defined in Annex L). The 'fixed-point IDCT' function indicates that an approximate inverse DCT (IDCT) may be used rather than the 'exact' definition of the IDCT given in the standard: this can be useful for low-complexity fixed-point implementations of the standard. The 'picture message' function allows the insertion of a user-definable message into the coded bit stream.

### 5.4.1   H.263 Profiles

It is very unlikely that all 19 optional modes will be required for any one application. Instead, certain combinations of modes may be useful for particular transmission scenarios. In common with MPEG-2 and MPEG-4, H.263 defines a set of recommended *profiles* (where a profile is a subset of the optional tools) and *levels* (where a level sets a maximum value on certain coding parameters such as frame resolution, frame rate and bit rate). Profiles and levels are defined in the final annex of H.263, Annex X. There are a total of nine profiles, as follows.

**Profile 0, Baseline**   This is simply the baseline H.263 functionality, without any optional modes.

**Profile 1, Coding efficiency** (Version 2)   This profile provides efficient coding using only tools available in Versions 1 and 2 of the standard (i.e. up to Annex T). The selected optional modes are Annex I (Advanced Intra-coding), Annex J (De-blocking Filter), Annex L (Supplemental Information: only the full picture freeze function is supported) and Annex T (Modified Quantisation). Annexes I, J and T provide improved coding efficiency compared with the baseline mode. Annex J incorporates the 'best' features of the first version of the standard, four motion vectors per macroblock and unrestricted motion vectors.

**Profile 2, Coding efficiency** (Version 1)    Only tools available in Version 1 of the standard are used in this profile and in fact only Annex F (Advanced Prediction) is included. The other three annexes (D, E, G) from the original standard are not (with hindsight) considered to offer sufficient coding gains to warrant their use.

**Profiles 3 and 4, Interactive and streaming wireless**    These profiles incorporate efficient coding tools (Annexes I, J and T) together with the slice structured mode (Annex K) and, in the case of Profile 4, the data partitioned slice mode (Annex V). These slice modes can support increased error resilience which is important for 'noisy' wireless transmission environments.

**Profiles 5, 6, 7, Conversational**    These three profiles support low-delay, high-compression 'conversational' applications (such as video telephony). Profile 5 includes tools that provide efficient coding; Profile 6 adds the slice structured mode (Annex K) for Internet conferencing; Profile 7 adds support for interlaced camera sources (part of Annex W).

**Profile 8, High latency**    For applications that can tolerate a higher latency (delay), such as streaming video, Profile 8 adds further efficient coding tools such as B-pictures (Annex O) and reference picture resampling (Annex P). B-pictures increase coding efficiency at the expense of a greater delay.

The remaining tools within the 19 annexes are not included in any profile, either because they are considered to be too complex for anything other than special-purpose applications, or because more efficient tools have superseded them.

## 5.5   H.26L[3]

The 19 optional modes of H.263 improved coding efficiency and transmission capabilities: however, development of H.263 standard is constrained by the requirement to continue to support the original 'baseline' syntax. The latest standardisation effort by the Video Coding Experts Group is to develop a new coding syntax that offers significant benefits over the older H.261 and H.263 standards. This new standard is currently described as 'H.26L', where the L stands for 'long term' and refers to the fact that this standard was planned as a long-term solution beyond the 'near-term' additions to H.263 (Versions 2 and 3).

The aim of H.26L is to provide a 'next generation' solution for video coding applications offering significantly improved coding efficiency whilst reducing the 'clutter' of the many optional modes in H.263. The new standard also aims to take account of the changing nature of video coding applications. Early applications of H.261 used dedicated CODEC hardware over the low-delay, low-error-rate ISDN. The recent trend is towards software-only or mixed software/hardware CODECs (where computational resources are limited, but greater flexibility is possible than with a dedicated hardware CODEC) and more challenging transmission scenarios (such as wireless links with high error rates and packet-based transmission over the Internet).

H.26L is currently at the test model development stage and may continue to evolve before standardisation. The main features can be summarised as follows.

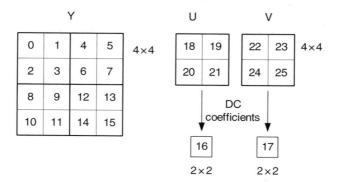

**Figure 5.7** H.26L blocks in a macroblock

**Processing units**   The basic unit is the macroblock, as with the previous standards. However, the subunit is now a $4 \times 4$ block (rather than an $8 \times 8$ block). A macroblock contains 26 blocks in total (Figure 5.7): 16 blocks for the luminance (each $4 \times 4$), four $4 \times 4$ blocks each for the chrominance components and two $2 \times 2$ 'sub-blocks' which hold the DC coefficients of each of the eight chrominance blocks. It is more efficient to code these DC coefficients together because they are likely to be highly correlated.

**Intra-prediction**   Before coding a $4 \times 4$ block within an intra-macroblock, each pixel in the block is predicted from previously coded pixels. This prediction reduces the amount of data coded in low-detail areas of the picture.

**Prediction reference for inter-coding**   In a similar way to Annexes N and U of H.263, the reference frame for predicting the current inter-coded macroblock may be selected from a range of previously coded frames. This can improve coding efficiency and error resilience at the expense of increased complexity and storage.

**Sub-pixel motion vectors**   H.26L supports motion vectors with $\frac{1}{4}$ pixel and (optionally) $\frac{1}{8}$ pixel accuracy; $\frac{1}{4}$-pixel vectors can give an appreciable improvement in coding efficiency over $\frac{1}{2}$-pixel vectors (e.g. H.263, MPEG-4) and $\frac{1}{8}$-pixel vectors can give a small further improvement (at the expense of increased complexity).

**Motion vector options**   H.26L offers seven different options for allocating motion vectors within a macroblock, ranging from one vector per macroblock (Mode 1 in Figure 5.8) to an individual vector for each of the 16 luminance blocks (Mode 7 in Figure 5.8). This makes it possible to model the motion of irregular-shaped objects with reasonable accuracy. More motion vectors require extra bits to encode and transmit and so the encoder must balance the choice of motion vectors against coding efficiency.

**De-blocking filter**   The de-blocking filter defined in Annex J of H.263 significantly improves motion compensation efficiency because it improves the 'smoothness' of the reference frame used for motion compensation. H.26L includes an integral de-blocking filter that operates across the edges of the $4 \times 4$ blocks within each macroblock.

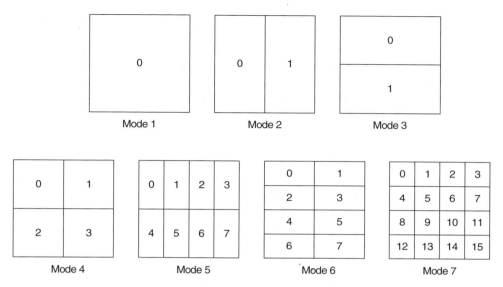

**Figure 5.8** H.26L motion vector modes

**4 × 4 Block transform**   After motion compensation, the residual data within each block is transformed using a 4 × 4 block transform. This is based on a 4 × 4 DCT but is an integer transform (rather than the floating-point 'true' DCT). An integer transform avoids problems caused by mismatches between different implementations of the DCT and is well suited to implementation in fixed-point arithmetic units (such as low-power embedded processors, Chapter 13).

**Universal variable-length code**   The VLC tables in H.263 are replaced with a single 'universal' VLC. A transmitted code is created by building up a regular VLC from the 'universal' codeword. These codes have two advantages: they can be implemented efficiently in software without the need for storage of large tables and they are reversible, making it easier to recover from transmission errors (see Chapters 8 and 11 for further discussion of VLCs and error resilience).

**Content-based adaptive binary arithmetic coding**   This alternative entropy encoder uses arithmetic coding (described in Chapter 8) to give higher compression efficiency than variable-length coding. In addition, the encoder can adapt to local image statistics, i.e. it can generate and use accurate probability statistics rather than using predefined probability tables.

**B-pictures**   These are recognised to be a very useful coding tool, particularly for applications that are not very sensitive to transmission delays. H.26L supports B-pictures in a similar way to MPEG-1 and MPEG-2, i.e. there is no restriction on the number of B-pictures that may be transmitted between pairs of I- and/or P-pictures.

At the time of writing it remains to be seen whether H.26L will supersede the popular H.261 and H.263 standards. Early indications are that it offers a reasonably impressive performance gain over H.263 (see the next section): whether these gains are sufficient to merit a 'switch' to the new standard is not yet clear.

## 5.6  PERFORMANCE OF THE VIDEO CODING STANDARDS

Each of the image and video coding standards described in Chapters 4 and 5 was designed for a different purpose and includes different features. This makes it difficult to compare them directly. Figure 5.9 compares the PSNR performance of each of the video coding standards for one particular test video sequence, 'Foreman', encoded at QCIF resolution and a frame rate of 10 frames per second. The results shown in the figure should be interpreted with caution, since different performance will be measured depending on the video sequence, frame rate and so on. However, the trend in performance is clear. MJPEG performs poorly (i.e. it requires a relatively high data rate to support a given picture 'quality') because it does not use any inter-frame compression. H.261 achieves a substantial gain over MJPEG, due to the use of integer-pixel motion compensation. MPEG-2 (with half-pixel motion compensation) is next, followed by H.263/MPEG-4 (which achieve a further gain by using four motion vectors per macroblock). The emerging H.26L test model achieves the best performance of all. (Note that MPEG-1 achieves the same performance as MPEG-2 in this test because the video sequence is not interlaced.)

This comparison is not the complete picture because it does not take into account the special features of particular standards (for example, the content-based tools of MPEG-4 or the interlaced video tools of MPEG-2). Table 5.1 compares the standards in terms of coding performance and features. At the present time, MPEG-2, H.263 and MPEG-4 are each viable

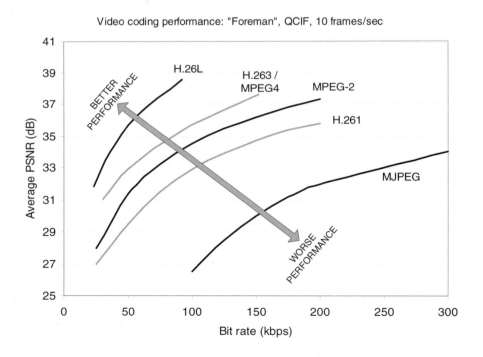

**Figure 5.9**  Coding performance comparison

**Table 5.1**   Comparison of the video coding standards

| Standard | Target application | Coding performance | Features |
|---|---|---|---|
| MJPEG | Image coding | 1 (worst) | Scalable and lossless coding modes |
| H.261 | Video conferencing | 2 | Integer-pixel motion compensation |
| MPEG-1 | Video-CD | 3 (equal) | I, P, B-pictures, half-pixel compensation |
| MPEG-2 | Digital TV | 3 (equal) | As above; field coding, scalable coding |
| H.263 | Video conferencing | 4 (equal) | Optimised for low bit rates; many optional modes |
| MPEG-4 | Multimedia coding | 4 (equal) | Many options including content-based tools |
| H.26L | Video conferencing | 5 (best) | Full feature set not yet defined |

alternatives for designers of video communication systems. MPEG-2 is a relatively mature technology for the mass-market digital television applications; H.263 offers good coding performance and options to support a range of transmission scenarios; MPEG-4 provides a large toolkit with the potential for new and innovative content-based applications. The emerging H.26L standard promises to outperform the H.263 and MPEG-4 standards in terms of video compression efficiency[4] but is not yet finalised.

## 5.7   SUMMARY

The ITU-T Video Coding Experts Group developed the H.261 standard for video conferencing applications which offered reasonable compression performance with relatively low complexity. This was superseded by the popular H.263 standard, offering better performance through features such as half-pixel motion compensation and improved variable-length coding. Two further versions of H.263 have been released, each offering additional optional coding modes to support better compression efficiency and greater flexibility. The latest version (Version 3) includes 19 optional modes, but is constrained by the requirement to support the original, 'baseline' H.263 CODEC. The H.26L standard, under development at the time of writing, incorporates a number of new coding tools such as a $4 \times 4$ block transform and flexible motion vector options and promises to outperform earlier standards.

Comparing the performance of the various coding standards is difficult because a direct 'rate-distortion' comparison does not take into account other factors such as features, flexibility and market penetration. It seems clear that the H.263, MPEG-2 and MPEG-4 standards each have their advantages for designers of video communication systems. Each of these standards makes use of common coding technologies: motion estimation and compensation, block transformation and entropy coding. In the next section of this book we will examine these core technologies in detail.

# REFERENCES

1. ITU-T Recommendation H.261, 'Video CODEC for audiovisual services at px64 kbit/s', 1993.
2. ITU-T Recommendation H.263, 'Video coding for low bit rate communication', Version 2, 1998.
3. ITU-T Q6/SG16 VCEG-L45, 'H.26L Test Model Long Term Number 6 (TML-6) draft 0', March 2001.
4. ITU-T Q6/SG16 VCEG-M08, 'Objective coding performance of [H.26L] TML 5.9 and H.263+', March 2001.

# 6

# Motion Estimation and Compensation

## 6.1  INTRODUCTION

In the video coding standards described in Chapters 4 and 5, blocks of image samples or residual data are compressed using a block-based transform (such as the DCT) followed by quantisation and entropy encoding. There is limited scope for improved compression performance in the later stages of encoding (DCT, quantisation and entropy coding), since the operation of the DCT and the codebook for entropy coding are specified by the relevant video coding standard. However, there is scope for significant performance improvement in the design of the first stage of a video CODEC (motion estimation and compensation). Efficient motion estimation reduces the energy in the motion-compensated residual frame and can dramatically improve compression performance. Motion estimation can be very computationally intensive and so this compression performance may be at the expense of high computational complexity. This chapter describes the motion estimation and compensation process in detail and discusses implementation alternatives and trade-offs.

The motion estimation and compensation functions have many implications for CODEC performance. Key performance issues include:

- Coding performance (how efficient is the algorithm at minimising the residual frame?)

- Complexity (does the algorithm make effective use of computation resources, how easy is it to implement in software or hardware?)

- Storage and/or delay (does the algorithm introduce extra delay and/or require storage of multiple frames?)

- 'Side' information (how much extra information, e.g. motion vectors, needs to be transmitted to the decoder?)

- Error resilience (how does the decoder perform when errors occur during transmission?)

These issues are interrelated and are potentially contradictory (e.g. better coding performance may lead to increased complexity and delay and poor error resilience) and different solutions are appropriate for different platforms and applications. The design and implementation of motion estimation, compensation and reconstruction can be critical to the performance of a video coding application.

## 6.2 MOTION ESTIMATION AND COMPENSATION

### 6.2.1 Requirements for Motion Estimation and Compensation

Motion estimation creates a *model* of the current frame based on available data in one or more previously encoded frames ('reference frames'). These reference frames may be 'past' frames (i.e. earlier than the current frame in temporal order) or 'future' frames (i.e. later in temporal order). The design goals for a motion estimation algorithm are to model the current frame as accurately as possible (since this gives better compression performance) whilst maintaining acceptable computational complexity. In Figure 6.1, the motion estimation module creates a model by modifying one or more reference frames to match the current frame as closely as possible (according to a matching criterion). The current frame is *motion compensated* by subtracting the model from the frame to produce a motion-compensated residual frame. This is coded and transmitted, along with the information required for the decoder to recreate the model (typically a set of *motion vectors*). At the same time, the encoded residual is decoded and added to the model to *reconstruct* a decoded copy of the current frame (which may not be identical to the original frame because of coding losses). This reconstructed frame is stored to be used as a reference frame for further predictions.

The residual frame (or displaced frame difference, DFD) is encoded and transmitted, together with any 'side information' (such as motion vectors) needed to recreate the model at the decoder. The 'best' compression performance is achieved when the size of the coded DFD and coded side information is minimised. The size of the coded DFD is related to the energy remaining in the DFD after motion compensation. Figure 6.2 shows a previous, current and residual frame (DFD) without motion compensation: there is clearly a significant amount of energy present around the boundaries of moving objects (the girl and the bicycle in this case). It should be possible to reduce this energy (and improve compression performance) using motion estimation and compensation.

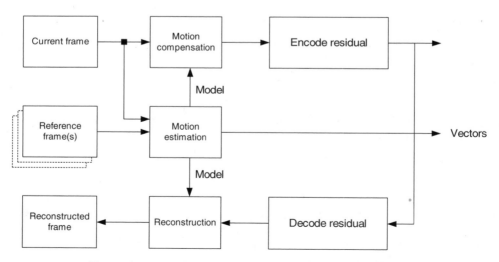

**Figure 6.1**  Motion estimation and compensation block diagram

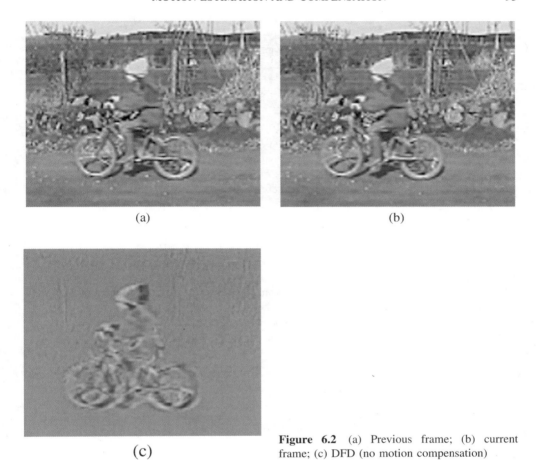

(a)  (b)

(c)

**Figure 6.2** (a) Previous frame; (b) current frame; (c) DFD (no motion compensation)

## 6.2.2 Block Matching

In the popular video coding standards (H.261, H.263, MPEG-1, MPEG-2 and MPEG-4), motion estimation and compensation are carried out on $8 \times 8$ or $16 \times 16$ blocks in the current frame. Motion estimation of complete blocks is known as *block matching*.

For each block of luminance samples (say $16 \times 16$) in the current frame, the motion estimation algorithm searches a neighbouring area of the reference frame for a 'matching' $16 \times 16$ area. The best match is the one that minimises the energy of the difference between the current $16 \times 16$ block and the matching $16 \times 16$ area. The area in which the search is carried out may be centred around the position of the current $16 \times 16$ block, because (a) there is likely to be a good match in the immediate area of the current block due to the high similarity (correlation) between subsequent frames and (b) it would be computationally intensive to search the whole of the reference frame.

Figure 6.3 illustrates the block matching process. The current 'block' (in this case, $3 \times 3$ pixels) is shown on the left and this block is compared with the same position in the reference frame (shown by the thick line in the centre) and the immediate neighbouring positions ($+/-1$ pixel in each direction). The mean squared error (MSE) between the

Current block

| | | |
|---|---|---|
| 1 | 3 | 2 |
| 6 | 4 | 3 |
| 5 | 4 | 3 |

Reference area

| | | | | |
|---|---|---|---|---|
| 1 | 3 | 2 | 4 | 5 |
| 6 | 4 | 2 | 3 | 2 |
| 5 | 4 | 2 | 2 | 3 |
| 4 | 4 | 3 | 3 | 1 |
| 4 | 6 | 7 | 4 | 5 |

Positions (x,y)

| | | |
|---|---|---|
| (-1,1) | (0,1) | (1,1) |
| (-1,0) | (0,0) | (1,0) |
| (-1,-1) | (0,-1) | (1,-1) |

**Figure 6.3** Current $3 \times 3$ block and $5 \times 5$ reference area

current block and the same position in the reference frame (position (0, 0)) is given by

$$\{(1-4)^2 + (3-2)^2 + (2-3)^2 + (6-4)^2 + (4-2)^2 + (3-2)^2 + (5-4)^2 + (4-3)^2 + (3-3)^2\}/9 = 2.44$$

The complete set of MSE values for each search position is listed in Table 6.1 and shown graphically in Figure 6.4. Of the nine candidate positions, $(-1, 1)$ gives the smallest MSE and hence the 'best' match. In this example, the best 'model' for the current block (i.e. the best prediction) is the $3 \times 3$ region in position $(-1, 1)$.

A video encoder carries out this process for each block in the current frame

1. Calculate the energy of the difference between the current block and a set of neighbouring regions in the reference frame.
2. Select the region that gives the lowest error (the 'matching region').
3. Subtract the matching region from the current block to produce a difference block.
4. Encode and transmit the difference block.
5. Encode and transmit a 'motion vector' that indicates the position of the matching region, relative to the current block position (in the above example, the motion vector is $(-1, 1)$.

Steps 1 and 2 above correspond to *motion estimation* and step 3 to *motion compensation*.

The video decoder reconstructs the block as follows:

1. Decode the difference block and motion vector.
2. Add the difference block to the matching region in the reference frame (i.e. the region 'pointed' to by the motion vector).

**Table 6.1** MSE values for block matching example

| Position $(x, y)$ | $(-1, -1)$ | $(0, -1)$ | $(1, -1)$ | $(-1, 0)$ | $(0, 0)$ | $(1, 0)$ | $(-1, 1)$ | $(0, 1)$ | $(1, 1)$ |
|---|---|---|---|---|---|---|---|---|---|
| MSE | 4.67 | 2.89 | 2.78 | 3.22 | 2.44 | 3.33 | 0.22 | 2.56 | 5.33 |

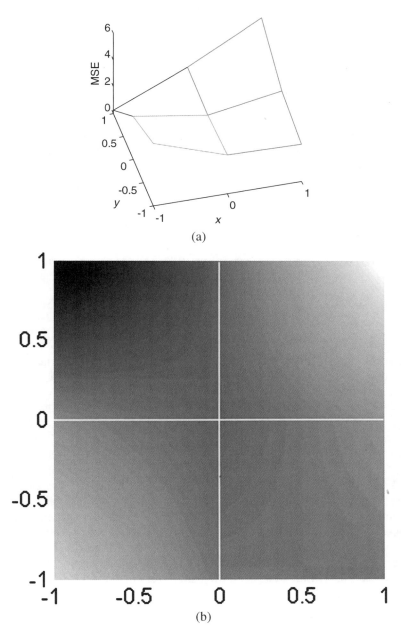

**Figure 6.4** MSE map: (a) surface plot; (b) pseudocolour plot

## 6.2.3 Minimising Difference Energy

The name 'motion estimation' is misleading because the process does not necessarily identify 'true' motion, instead it attempts to find a matching region in the reference frame

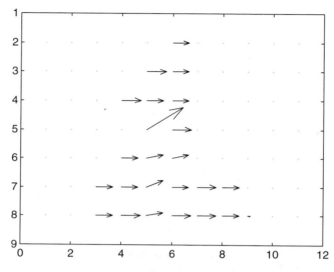

**Figure 6.5**   $16 \times 16$ block motion vectors

that minimises the energy of the difference block. Where there is clearly identifiable linear motion, such as large moving objects or global motion (camera panning, etc.), motion vectors produced in this way should roughly correspond to the movement of blocks between the reference and the current frames. However, where the motion is less obvious (e.g. small moving objects that do not correspond to complete blocks, irregular motion, etc.), the 'motion vector' may not indicate genuine motion but rather the position of a good match.

Figure 6.5 shows the motion vectors produced by motion estimation for each of the $16 \times 16$ blocks ('macroblocks') of the frame in Figure 6.2. Most of the vectors do correspond to motion: the girl and bicycle are moving to the *left* and so the vectors point to the *right* (i.e. to the region the objects have moved *from*). There is an anomalous vector in the middle (it is larger than the rest and points diagonally upwards). This vector does not correspond to 'true' motion, it simply indicates that the best match can be found in this position.

There are many possible variations on the basic block matching process, some of which will be described later in this chapter. Alternative measures of DFD energy may be used (to reduce the computation required to calculate MSE). Varying block sizes, or irregular-shaped regions, can be more efficient at matching 'true' motion than fixed $16 \times 16$ blocks. A better match may be found by searching within two or more reference frames (rather than just one). The order of searching neighbouring regions can have a significant effect on matching efficiency and computational complexity. Objects do not necessarily move by an integral number of pixels between successive frames and so a better match may be obtained by searching sub-pixel positions in the reference frame. The block matching process itself only works well for large, regular objects with linear motion: irregular objects and non-linear motion (such as rotation or deformation) may be modelled more accurately with other motion estimation methods such as object-based or mesh-based estimation.

*Comparison criteria*

Mean squared error provides a measure of the energy remaining in the difference block. MSE for a $N \times N$-sample block can be calculated as follows:

$$\text{MSE} = \frac{1}{N^2} \sum_{i=0}^{N-1} \sum_{j=0}^{N-1} (C_{ij} - R_{ij})^2 \tag{6.1}$$

where $C_{ij}$ is a sample of the current block, $R_{ij}$ is a sample of the reference area and $C_{00}$, $R_{00}$ are the top-left samples in the current and reference areas respectively.

Mean absolute error (MAE) provides a reasonably good approximation of residual energy and is easier to calculate than MSE, since it requires a magnitude calculation instead of a square calculation for each pair of samples:

$$\text{MAE} = \frac{1}{N^2} \sum_{i=0}^{N-1} \sum_{j=0}^{N-1} |C_{ij} - R_{ij}| \tag{6.2}$$

The comparison may be simplified further by neglecting the term $1/N^2$ and simply calculating the sum of absolute errors (SAE) or sum of absolute differences (SAD):

$$\text{SAE} = \sum_{i=0}^{N-1} \sum_{j=0}^{N-1} |C_{ij} - R_{ij}| \tag{6.3}$$

SAE gives a reasonable approximation to block energy and so Equation 6.3 is a commonly used matching criterion for block-based motion estimation.

## 6.3 FULL SEARCH MOTION ESTIMATION

In order to find the best matching region in the reference frame, in theory it is necessary to carry out a comparison of the current block with every possible region of the reference frame. This is usually impractical because of the large number of comparisons required. In practice, a good match for the current block can usually be found in the immediate neighbourhood of the block position in the reference frame (if a match exists). Hence in practical implementations, the search for a matching region is limited to a 'search window', typically centred on the current block position.

The optimum size of search window depends on several factors: the resolution of each frame (a larger window is appropriate for a higher resolution), the type of scene (high-motion scenes benefit from a larger search window than low-motion scenes) and the

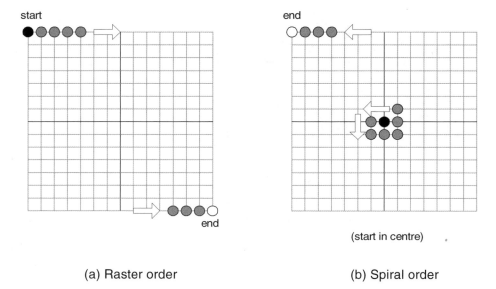

(a) Raster order                    (b) Spiral order

**Figure 6.6**  Full search: (a) raster and (b) spiral order

available processing resources (since a larger search window requires more comparison operations and hence more processing).

*Full search motion estimation* calculates the comparison criterion (such as SAE) at each possible location in the search window. Full search is computationally intensive, particularly for large search windows. The locations may be processed in a 'raster' order (Figure 6.6, left-hand diagram) or in a spiral order starting from the centre (0, 0) position (Figure 6.6, right-hand diagram). The spiral search order has certain computational advantages when early termination algorithms are used (see Section 6.9.1) because the best match (and hence the smallest SAE) is most likely to occur near the centre of the search region.

Figure 6.7 shows an example of the SAE results for a full search. The figure shows the current block and the reference area ($+/-15$ pixels around the current $16 \times 16$ block position) together with a plot of the SAE values found at each search location. There are a total of $31 \times 31$ SAE values (corresponding to integer steps from $-15$ to $+15$ in the $x$ and $y$ directions). The smallest SAE value can be found at location ($x = 6$, $y = 1$) and is marked on the SAE plot. This is the global minimum of the SAE function in the search region and the full search algorithm will select this position as the 'best' match. Note that there are other, local minima of the SAE function (the dark 'patches' on the SAE plot): the importance of these local minima will become clear in the next section.

The effect of motion estimation and compensation is illustrated in Figure 6.8. After motion estimation (using full search block matching) and compensation, the reference frame (shown in Figure 6.2) is 'reorganised' to provide a closer match to the current frame. The motion-compensated DFD shown in Figure 6.8 contains less energy than the uncompensated DFD in Figure 6.2 and will therefore produce a smaller coded frame.

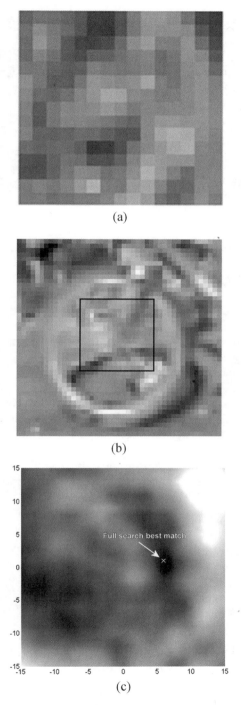

**Figure 6.7**  SAD map: (a) current block; (b) search area; (c) map with minima

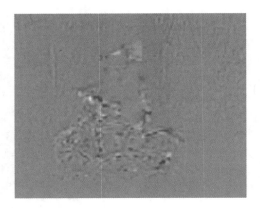

**Figure 6.8**   Residual frame after full search
motion estimation and compensation

## 6.4   FAST SEARCH

The computational complexity of a full search algorithm is often prohibitive, particularly for
software CODECs that must operate in 'real time'. Many alternative 'fast search' algorithms
have been developed for motion estimation. A fast search algorithm aims to reduce the
number of comparison operations compared with full search, i.e. a fast search algorithm will
'sample' just a few of the points in the SAE map whilst attempting to find the minimum
SAE. The critical question is whether the fast algorithm can locate the 'true' minimum rather
than a 'local' minimum. Whereas the full search algorithm is guaranteed to find the global
minimum SAE, a search algorithm that samples only some of the possible locations in the
search region may get 'trapped' in a local minimum. The result is that the difference block
found by the fast search algorithm contains more energy than the block found by full search
and hence the number of coded bits generated by the video encoder will be larger.
Because of this, fast search algorithms usually give poorer compression performance than
full search.

### 6.4.1   Three-Step Search (TSS)[1]

This algorithm is most widely known in its three-step form, the 'three-step search' (TSS),
but it can be carried out with other numbers of steps (i.e. $N$-step search). For a search
window of $+/-(2^N - 1)$ pixels, the TSS algorithm is as follows:

1. Search location (0, 0).
2. Set $S = 2^{N-1}$ (the step size).
3. Search eight locations $+/-S$ pixels around location (0, 0).
4. From the nine locations searched so far, pick the location with the smallest SAE and
   make this the new search origin.

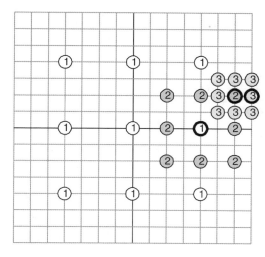

**Figure 6.9**   Three-step search (TSS)

5. Set $S = S/2$.

6. Repeat stages 3–5 until $S = 1$.

Figure 6.9 illustrates the procedure for a search window of $+/-7$ (i.e. $N = 3$). The first 'step' involves searching location $(0, 0)$ and eight locations $+/-4$ pixels around the origin. The second 'step' searches $+/-2$ pixels around the best match from the first step (highlighted in bold) and the third step searches $+/-1$ pixels around the best match from the second step (again highlighted). The best match from this third step is chosen as the result of the search algorithm. With a search window of $+/-7$, three repetitions (steps) are required to find the best match. A total of $(9+8+8) = 25$ search comparisons are required for the TSS, compared with $(15 \times 15) = 225$ comparisons for the equivalent full search. In general, $(8N+1)$ comparisons are required for a search area of $+/-(2^N - 1)$ pixels.

## 6.4.2   Logarithmic Search[2]

The logarithmic search algorithm can be summarised as follows:

1. Search location $(0, 0)$.

2. Search four locations in the horizontal and vertical directions, $S$ pixels away from the origin (where $S$ is the initial step size). The five locations make a '$+$' shape.

3. Set the new origin to the best match (of the five locations tested). If the best match is at the centre of the '$+$', $S = S/2$, otherwise $S$ is unchanged.

4. If $S = 1$ then go to stage 5, otherwise go to stage 2.

5. Search eight locations immediately surrounding the best match. The search result is the best match of the search origin and these eight neighbouring locations.

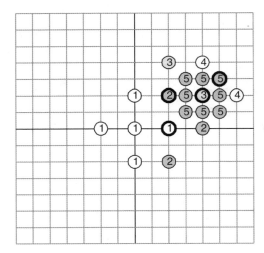

**Figure 6.10**  Logarithmic search

Figure 6.10 shows an example of the search pattern with $S = 2$ initially. Again, the best match at each iteration is highlighted in bold (note that the bold 3 is the best match at iteration 3 *and* at iteration 4). In this example 20 search comparisons are required: however, the number of comparisons varies depending on number of repetitions of stages 2, 3 and 4 above. Note that the algorithm will not search a candidate position if it is outside the search window ($+/-7$ in this example).

### 6.4.3   Cross-Search[3]

This algorithm is similar to the three-step search except that five points are compared at each step (forming an X) instead of nine.

1. Search location (0, 0).

2. Search four locations at $+/-S$, forming an 'X' shape (where $S = 2^{N-1}$ as for the TSS).

3. Set the new origin to be the best match of the five locations tested.

4. If $S > 1$ then $S = S/2$ and go to stage 2; otherwise go to stage 5.

5. If the best match is at the top left or bottom right of the 'X', evaluate four more points in an 'X' at a distance of $+/-1$; otherwise (best match is at the top right or bottom left) evaluate four more points in a '+' at a distance of $+/-1$.

Figure 6.11 shows two examples of the cross-search algorithm: in the first example, the final points are in the shape of a 'X' and in the second, they are in the shape of a '+' (the best match at each iteration is highlighted). The number of SAD comparisons is $(4N + 5)$ for a search area of $+/-(2^N - 1)$ pixels (i.e. 17 comparisons for a $+/-7$ pixel window).

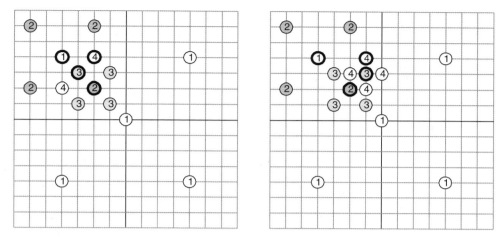

**Figure 6.11**   Cross-search

### 6.4.4   One-at-a-Time Search

This simple algorithm essentially involves following the SAD 'gradient' in the horizontal direction until a minimum is found, then following the gradient in the vertical direction to find a vertical minimum:

1. Set the horizontal origin to (0, 0).

2. Search the origin and the two immediate horizontal neighbours.

3. If the origin has the smallest SAD (of the three neighbouring horizontal points), then go to stage 5, otherwise. . . .

4. Set the new origin to the horizontal point with the smallest SAD and search the neighbouring point that has not yet been searched. Go to stage 3.

5. Repeat stages 2–4 in the vertical direction.

The one-at-a-time search is illustrated in Figure 6.12. The positions marked 1 are searched and the left-hand position gives the best match. Position 2 is searched and gives the best match. The horizontal search continues with positions 3, 4 and 5 until position 4 is found to have a lower SAD than position 5 (i.e. a horizontal minimum has been detected). The vertical search starts with positions 6: the best match is at the top and the vertical search continues with 7, 8, 9 until a minimum is detected at position 8. In this example only nine searches are carried out: however, there is clearly potential to be trapped in a local minimum.

### 6.4.5   Nearest Neighbours Search[4]

This algorithm was proposed for H.263 and MPEG-4 (short header) CODECs. In these CODECs, each motion vector is predicted from neighbouring (already coded) motion vectors prior to encoding (see Figure 8.3). This makes it preferable to choose a vector close to this

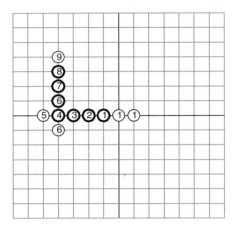

**Figure 6.12**  One-at-a-time search

'median predictor' position, for two reasons. First, neighbouring macroblocks often have similar motion vectors (so that there is a good chance that the median predictor will be close to the 'true' best match). Second, a vector near the median will have a small displacement and therefore a small VLC.

The algorithm proceeds as follows:

1. Search the (0, 0) location.

2. Set the search origin to the predicted vector location and search this position.

3. Search the four neighbouring positions to the origin in a '+' shape.

4. If the search origin (or location 0, 0 for the first iteration) gives the best match, this is the chosen search result; otherwise, set the new origin to the position of the best match and go to stage 3.

The algorithm stops when the best match is at the centre of the '+' shape (or the edge of the search window has been reached). An example of a search sequence is shown in Figure 6.13.

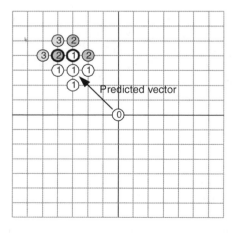

**Figure 6.13**  Nearest neighbours search

The median predicted vector is $(-3, 3)$ and this is shown with an arrow. The $(0, 0)$ point (marked 0) and the first 'layer' of positions (marked 1) are searched: the best match is highlighted. The layer 2 positions are searched, followed by layer 3. The best match for layer 3 is in the centre of the '+' shape and so the search is terminated.

This algorithm will perform well if the motion vectors are reasonably homogeneous, i.e. there are not too many sudden changes in the motion vector field. The algorithm described in [4] includes two further features. First, if the median predictor is unlikely to be accurate (because too many neighbouring macroblocks are intra-coded and therefore have no motion vectors), an alternative algorithm such as the TSS is used. Second, a cost function is proposed to estimate whether the computational complexity of carrying out the next set of searches is worthwhile. (This will be discussed further in Chapter 10.)

## 6.4.6 Hierarchical Search

The hierarchical search algorithm (and its variants) searches a coarsely subsampled version of the image first, followed by successively higher-resolution versions until the full image resolution is reached:

1. Level 0 consists of the current and reference frames at their full resolutions. Subsample level 0 by a factor of 2 in the horizontal and vertical directions to produce level 1.

2. Repeat, subsampling level 1 to produce level 2, and so on until the required number of levels are available (typically, three or four levels are sufficient).

3. Search the highest level to find the best match: this is the initial 'coarse' motion vector.

4. Search the next lower level around the position of the 'coarse' motion vector and find the best match.

5. Repeat stage 4 until the best match is found at level 0.

The search method used at the highest level may be full search or a 'fast' algorithm such as TSS. Typically, at each lower level only $+/-1$ pixels are searched around the coarse vector. Figure 6.14 illustrates the method with three levels (2, 1 and 0) and a window of $+/-3$ positions at the highest level. A full search is carried out at the top level: however, the complexity is relatively low because we are only comparing a $4 \times 4$ pixel area at each level 2 search location. The best match (the number '2') is used as the centre of the level 1 search, where eight surrounding locations are searched. The best match (number '1') is used as the centre of the final level 0 search. The equivalent search window is $+/-15$ pixels (i.e. the algorithm can find a match anywhere within $+/-15$ pixels of the origin at level 0).

In total, 49 searches are carried out at level 2 (each comparing $4 \times 4$ pixel regions), 8 searches at level 1 (each comparing $8 \times 8$ pixel regions) and 8 searches at level 0 (comparing $16 \times 16$ pixel regions).

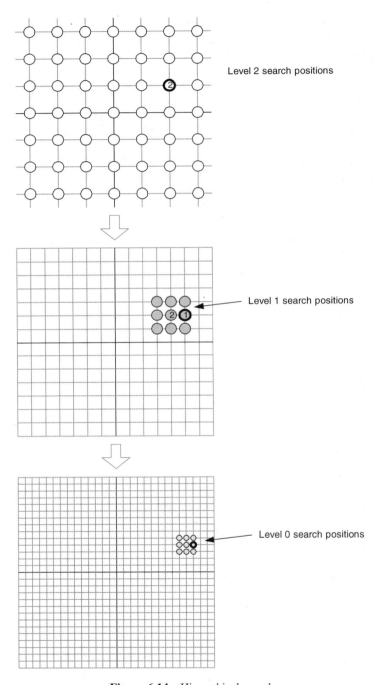

**Figure 6.14** Hierarchical search

## 6.5   COMPARISON OF MOTION ESTIMATION ALGORITHMS

The wide range of algorithms available for block-based motion estimation can make it difficult to choose between them. There are a number of criteria that may help in the choice:

1. Matching performance: how effective is the algorithm at minimising the residual block?

2. Rate-distortion performance: how well does the complete CODEC perform at various compressed bit rates?

3. Complexity: how many operations are required to complete the matching process?

4. Scalability: does the algorithm perform equally well for large and small search windows?

5. Implementation: is the algorithm suitable for software or hardware implementation for the chosen platform or architecture?

Criteria 1 and 2 appear to be identical. If the algorithm is effective at minimising the energy in the motion-compensated residual block, then it ought to provide good compression efficiency (good image quality at a low compressed bit rate). However, there are other factors that complicate things: for example, every motion vector that is calculated by the motion estimation algorithm must be encoded and transmitted as part of the compressed bit stream. As will be discussed in Chapter 8, larger motion vectors are usually coded with more bits and so an algorithm that efficiently minimises the residual frame but produces large motion vectors may be less efficient than an algorithm that is 'biased' towards producing small motion vectors.

*Example*

In the following example, block-based motion estimation and compensation were carried out on five frames of the 'bicycle' sequence (shown in Figure 6.2). Table 6.2 compares the performance of full search motion estimation with a range of search window sizes. The table lists the total SAE of the five difference frames without motion compensation (i.e. simply subtracting the previous from the current frame) and with motion compensation (i.e. block-based motion compensation on $16 \times 16$ blocks). The final column lists the total number of comparison operations (where one operation is the comparison of two luminance samples, $|C_{ij} - R_{ij}|$). As the search window increases, motion compensation efficiency improves (shown by a smaller SAE): however, the number of operations increases exponentially with the window size. This sequence contains relatively low movement and so most of the

**Table 6.2**   Full search motion estimation, five frames: varying window size

| Search window | Total SAE (uncompensated) | Total SAE (compensated) | Number of comparison operations |
|---|---|---|---|
| $+/-1$ | 1 326 783 | 1 278 610 | $1.0 \times 10^6$ |
| $+/-3$ | . . . | 1 173 060 | $5.2 \times 10^6$ |
| $+/-7$ | . . . | 898 581 | $23.4 \times 10^6$ |
| $+/-15$ | . . . | 897 163 | $99.1 \times 10^6$ |

**Table 6.3**  Motion estimation algorithm comparison, five frames: search window $= +/-15$

| Algorithm | Total SAE (uncompensated) | Total SAE (compensated) | Number of comparison operations |
|---|---|---|---|
| Full search | 1 326 783 | 897 163 | $99.1 \times 10^6$ |
| Three-step search | . . . | 914 753 | $3.6 \times 10^6$ |

performance gain from motion estimation is achieved with a search window of $+/-7$ samples. Increasing the window to $+/-15$ gives only a modest improvement in SAE at the expense of a fourfold increase in computation.

Table 6.3 compares the performance of full search and three-step search with a search window of $+/-15$ pixels. Full search produces a lower SAE and hence a smaller residual frame than TSS. However, the slight increase in SAE produced by the TSS algorithm is offset by a substantial reduction in the number of comparison operations.

Figure 6.15 shows how a fast search algorithm such as the TSS may fail to find the best possible match. The three-step search algorithm starts by considering the positions

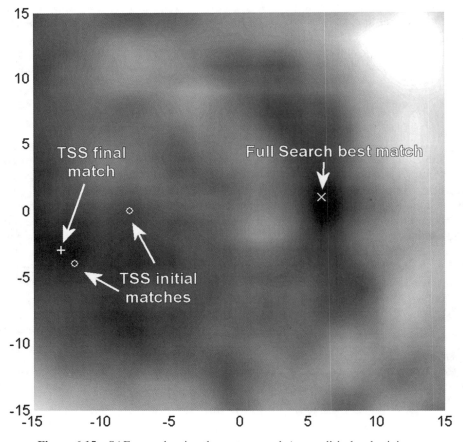

**Figure 6.15**  SAE map showing three-step search 'trapped' in local minimum

$+/-8$ pixels around the origin. The best match at the first step is found at $(-8, 0)$ and this is marked with a circle on the figure. The next step examines positions within $+/-4$ pixels of this point and the best of these is found at $(-12, -4)$. Step 3 also chooses the point $(-12, -4)$ and the final step selects $(-13, -3)$ as the best match (shown with a '$+$'). This point is a *local* minimum but not the *global* minimum. Hence the residual block after motion compensation will contain more energy than the best match found by the full search algorithm (point 6, 1 marked with an 'x').

Of the other search algorithms mentioned above, logarithmic search, cross-search and one-at-a-time search provide low computational complexity at the expense of relatively poor matching performance. Hierarchical search can give a good compromise between performance and complexity and is well suited to hardware implementations. Nearest-neighbours search, with its in-built 'bias' towards the median-predicted motion vector, is reported to perform almost as well as full search, with a very much reduced complexity. The high performance is achieved because the 'bias' tends to produce very small (and hence very efficiently coded) motion vectors and this efficiency offsets the slight drop in SAE performance.

## 6.6  SUB-PIXEL MOTION ESTIMATION

So far, we have assumed that the best match can be found at a region offset from the current block by an integer number of pixels. In fact, for many blocks a better match (and hence a smaller DFD) can be obtained by searching a region interpolated to sub-pixel accuracy. The search algorithm is extended as follows:

1. Interpolate between the samples of the search area in the reference frame to form a higher-resolution interpolated region.

2. Search full-pixel and sub-pixel locations in the interpolated region and find the best match.

3. Subtract the samples of the matching region (whether full- or sub-pixel) from the samples of the current block to form the difference block.

Half-pixel interpolation is illustrated in Figure 6.16. The original integer pixel positions 'a' are shown in black. Samples b and c (grey) are formed by linear interpolation between pairs

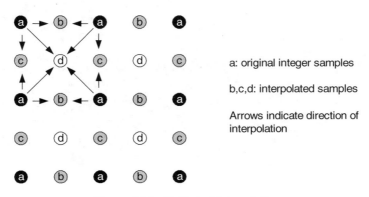

**Figure 6.16**  Half-pixel interpolation

of integer pixels, and samples d (white) are interpolated between four integer pixels (as indicated by the arrows). Motion compensation with half-pixel accuracy is supported by the H.263 standard, and higher levels of interpolation ($\frac{1}{4}$ pixel or more) are proposed for the emerging H.26L standard. Increasing the 'depth' of interpolation gives better block matching performance at the expense of increased computational complexity.

Searching on a sub-pixel grid obviously requires more computation than the integer searches described earlier. In order to limit the increase in complexity, it is common practice to find the best matching integer position and then to carry out a search at half-pixel locations immediately around this position. Despite the increased complexity, sub-pixel motion estimation and compensation can significantly outperform integer motion estimation/ compensation. This is because a moving object will not necessarily move by an integral number of pixels between successive video frames. Searching sub-pixel locations as well as integer locations is likely to find a good match in a larger number of cases.

Interpolating the reference area shown in Figure 6.7 to half-pixel accuracy and comparing the current block with each half-pixel position gives the SAE map shown in Figure 6.17. The best match (i.e. the lowest SAE) is found at position (6, 0.5). The block found at this position

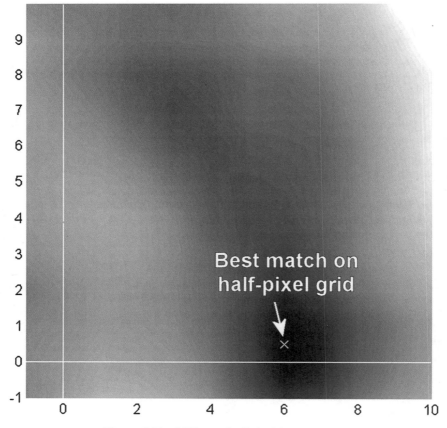

**Figure 6.17**  SAE map (half-pixel interpolation)

in the interpolated reference frame gives a better match than position (6, 1) and hence better motion compensation performance.

## 6.7 CHOICE OF REFERENCE FRAMES

The most 'obvious' choice of reference frame is the previous coded frame, since this should be reasonably similar to the current frame and is available in the encoder and decoder. However, there can be advantages in choosing from one or more other reference frames, either before or after the current frame in temporal order.

### 6.7.1 Forward Prediction

Forward prediction involves using an 'older' encoded frame (i.e. a preceding frame in temporal order) as prediction reference for the current frame. Forward prediction performs poorly in certain cases, for example:

1. when there is a significant time difference between the reference frame and the current frame (which may mean that the image has changed significantly);

2. when a scene change or 'cut' occurs;

3. when a moving object uncovers a previously hidden area of the image (e.g. a door opens): the hidden area does not exist in the reference frame and so cannot be efficiently predicted.

### 6.7.2 Backwards Prediction

The prediction efficiency for cases (2) and (3) above can be improved by using a 'future' frame (i.e. a later frame in temporal order) as prediction reference. A frame immediately after a scene cut, or an uncovered object, can be better predicted from a future frame.

Backwards prediction requires the encoder to buffer coded frames and encode them out of temporal order, so that the future reference frame is encoded before the current frame.

### 6.7.3 Bidirectional Prediction

In some cases, bidirectional prediction may outperform forward or backward prediction: here, the prediction reference is formed by 'merging' forward and backward references.

Forward, backward and bidirectional predictions are all available for encoding an MPEG-1 or MPEG-2 B-picture. Typically, the encoder carries out two motion estimation searches for each macroblock ($16 \times 16$ luminance samples), one based on the previous reference picture (an I- or P-picture) and one based on the future reference picture. The encoder finds the motion vector that gives the best match (i.e. the minimum SAE) based on (a) the previous reference frame and (b) the future reference frame. A third SAE value (c) is calculated by subtracting the average of the two matching areas (previous and future) from the current macroblock. The encoder chooses the 'mode' of the current macroblock based on the smallest of these three SAE values:

(a) forward prediction

(b) backwards prediction, or

(c) bidirectional prediction.

In this way, the encoder can find the optimum prediction reference for each macroblock and this improves compression efficiency by up to 50% for B-pictures.

### 6.7.4 Multiple Reference Frames

MPEG-1 or MPEG-2 B-pictures are encoded using two reference frames. This approach may be extended further by allowing the encoder to choose a reference frame from a large number of previously encoded frames. Choosing between multiple possible reference frames can be a useful tool in improving error resilience (as discussed in Chapter 11). This method is supported by the H.263 standard (Annexes N and U, see Chapter 5) and has been analysed in.[5]

Encoder and decoder complexity and storage requirements increase as more prediction reference frames are utilised. 'Simple' forward prediction from the previous encoded frame gives the lowest complexity (but also the poorest compression efficiency), whilst the other methods discussed above add complexity (and potentially encoding delay) but give improved compression efficiency.

Figure 6.18 illustrates the prediction options discussed above, showing forward and backwards prediction from past and future frames.

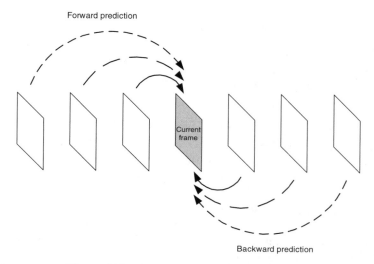

**Figure 6.18**    Reference frame prediction options

## 6.8 ENHANCEMENTS TO THE MOTION MODEL

Bidirectional prediction and multiple reference frames (described above) can increase compression efficiency because they improve the motion model, allowing a wider range of prediction options for each coded macroblock than a simple forward prediction from the previous encoded frame. Sub-pixel interpolation of the reference frame also improves the motion model by catering for the case when motion does not map neatly onto integer-pixel locations. There are a number of other ways in which the motion model may be enhanced, some of which are listed here.

### 6.8.1 Vectors That can Point Outside the Reference Picture

If movement occurs near the edges of the picture, the best match for an edge block may actually be offset slightly outside the boundaries of the reference picture. Figure 6.19 shows an example: the ball that has appeared in the current frame is partly visible in the reference frame and part of the best matching block will be found slightly above the boundary of the frame. The match may be improved by extrapolating the pixel values at the edge of the reference picture. Annex D of H.263 supports this type of prediction by simple linear extrapolation of the edge pixels into the area around the frame boundaries (shown in Figure 6.19). Block matching efficiency and hence compression efficiency is slightly improved for video sequences containing motion near the edges of the picture.

### 6.8.2 Variable Block Sizes

Using a block size of 16 × 16 for motion estimation and compensation gives a rather 'crude' model of image structure and motion. The advantages of a large block size are simplicity and the limited number of vectors that must be encoded and transmitted. However, in areas of complex spatial structure and motion, better performance can be achieved with smaller block sizes. H.263 Annex F enables an encoder to switch between a block size of 16 × 16 (one

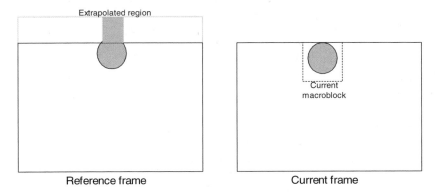

**Figure 6.19** Example of best match found outside the reference picture

motion vector per macroblock) and $8 \times 8$ (four vectors per macroblock) : the small block size is used when it gives better coding performance than the large block size. Motion compensation performance is noticeably improved at the expense of an increase in complexity: carrying out 4 searches per macroblock (albeit on a smaller block size with only 64 calculations per SAE comparison) requires more operations.

The emerging H.26L standard takes this approach further and supports multiple possible block sizes for motion compensation within a macroblock. Motion compensation may be carried out for sub-blocks with horizontal or vertical dimensions of any combination of 4, 8 or 16 samples. The extreme cases are $4 \times 4$ sub-blocks (resulting in 16 vectors per macroblock) and $16 \times 16$ blocks (one vector per macroblock) with many possibilities in between ($4 \times 8$, $8 \times 8$, $4 \times 16$ blocks, etc.). This flexibility gives a further increase in compression performance at the expense of higher complexity.

### 6.8.3   Overlapped Block Motion Compensation (OBMC)

When OBMC is used, each sample of the reference block used for motion compensation is formed by combining three predictions:

1. a sample predicted using the motion vector of the current block ($R_0$);

2. a sample predicted using the motion vector of the adjacent block in the vertical direction (i.e. the nearest neighbour block above or below) ($R_1$);

3. a sample predicted using the motion vector of the adjacent block in the horizontal direction (i.e. the nearest neighbour block left or right) ($R_2$).

The final sample is a weighted average of the three values. $R_0$ is given the most weight (because it uses the current block's motion vector). $R_1$ and $R_2$ are given more weight when the current sample is near the edge of the block, less weight when it is in the centre of the block.

The result of OBMC is to 'smooth' the prediction across block boundaries in the reference frame. OBMC is supported by Annex F of H.263 and gives a slight increase in motion compensation performance (at the expense of a significant increase in complexity). A similar 'smoothing' effect can be obtained by applying a filter to the block edges in the reference frame and later versions of H.263 (H.263+ and H.263++) recommend using a block filter instead of OBMC because it gives similar performance with lower computational complexity. OBMC and filtering performance have been discussed elsewhere,[6] and filters are examined in more detail in Chapter 9.

### 6.8.4   Complex Motion Models

The motion estimation and compensation schemes discussed so far have assumed a simple translational motion model, i.e. they work best when all movement in a scene occurs in a plane perpendicular to the viewer. Of course, there are many other types of movement such as rotation, movements towards or away from the viewer (zooming) and deformation of objects (such as a human body). Better motion compensation performance may be achieved by matching the current frame to a more complex motion model.

In the MPEG-4 standard, a video object plane may be predicted from the pixels that exist only within a reference VOP. This is a form of *region-based motion compensation*, where

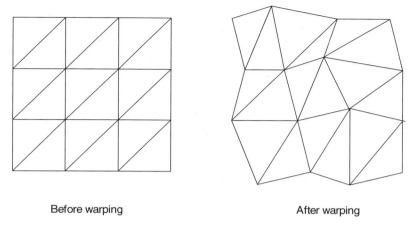

Before warping                                                    After warping

**Figure 6.20**   Triangular mesh before and after deformation

compensation is carried out on arbitrarily shaped regions rather than fixed rectangular blocks. This has the capability to provide a more accurate motion model for 'natural' video scenes (where moving objects rarely have 'neat' rectangular boundaries).

*Picture warping* involves applying a global warping transformation to the entire reference picture for example to compensate for global movements such as camera zoom or camera rotation.

*Mesh-based motion compensation* overlays the reference picture with a 2-D mesh of triangles. The motion-compensated reference is formed by moving the corners of each triangle and deforming the reference picture pixels accordingly (Figure 6.20 shows the general approach). A deformable mesh can model a wide range of movements, including object rotation, zooming and limited object deformations. A smaller mesh will give a more accurate motion model (but higher complexity).

Still more accurate modelling may be achieved using *object-based coding* where the encoder attempts to maintain a 3-D model of the video scene. Changes between frames are modelled by moving and deforming the components of the 3-D scene.

Picture warping is significantly more complex than 'standard' block matching. Mesh-based and object-based coding are successively more complex and are not suitable for real-time applications with current processing technology. However, they offer significant potential for future video coding systems when more processing power becomes available. These and other motion models are active areas for research.

## 6.9   IMPLEMENTATION

### 6.9.1   Software Implementations

Unless dedicated hardware assistance is available (e.g. a motion estimation co-processor), the key issue in a software implementation of motion estimation is the trade-off between computational complexity (the total number of processor cycles required) and compression performance. Other important considerations include:

- *The efficiency of the mapping to the target processor.* For example, an algorithm that fully utilises the instruction pipeline of the processor is preferable to an algorithm that introduces data dependencies and 'stalls' into the pipeline.

- *Data storage and delay requirements.* For example, there may be advantages to carrying out motion estimation for the entire frame before further encoding takes place: however, this requires more storage and can introduce more delay than an implementation where each macroblock is estimated, compensated, encoded and transmitted before moving onto the next macroblock.

Even with the use of fast search algorithms, motion estimation is often the most computationally intensive operation in a software video CODEC and so it is important to find ways to speed up the process. Possible approaches to optimising the code include:

1. *Loop unrolling.* Figure 6.21 lists pseudocode for two possible versions of the SAE calculation (Equation 6.3) for a $16 \times 16$ block. Version (a) is a direct, compact implementation of the equation. However, each of the $16 \times 16 = 256$ calculations is accompanied by incrementing and checking the inner loop counter $i$. Version (b) 'unrolls'

```
(a) Direct implementation:

// Current position: i,j Offset in reference frame: ioffset, joffset

totalSAE = 0;

for j = 0 to 15   {          // Row counter
      for i = 0 to 15   {        // Column counter
          totalSAE = totalSAE + abs(C[i,j] - R[i+ioffset,j+joffset]);
      }
}
```

```
(b) Unrolled inner loop:
// Current position: i,j Offset in reference frame: ioffset, joffset

totalSAE = 0;

for j = 0 to 15   {          // Row counter
      totalSAE = totalSAE + abs(C[0,j] - R[0+ioffset,j+joffset]);
      totalSAE = totalSAE + abs(C[1,j] - R[1+ioffset,j+joffset]);
      totalSAE = totalSAE + abs(C[2,j] - R[2+ioffset,j+joffset]);
      totalSAE = totalSAE + abs(C[3,j] - R[3+ioffset,j+joffset]);
      totalSAE = totalSAE + abs(C[4,j] - R[4+ioffset,j+joffset]);
      totalSAE = totalSAE + abs(C[5,j] - R[5+ioffset,j+joffset]);
      totalSAE = totalSAE + abs(C[6,j] - R[6+ioffset,j+joffset]);
      totalSAE = totalSAE + abs(C[7,j] - R[7+ioffset,j+joffset]);
      totalSAE = totalSAE + abs(C[8,j] - R[8+ioffset,j+joffset]);
      totalSAE = totalSAE + abs(C[9,j] - R[9+ioffset,j+joffset]);
      totalSAE = totalSAE + abs(C[10,j] - R[10+ioffset,j+joffset]);
      totalSAE = totalSAE + abs(C[11,j] - R[11+ioffset,j+joffset]);
      totalSAE = totalSAE + abs(C[12,j] - R[12+ioffset,j+joffset]);
      totalSAE = totalSAE + abs(C[13,j] - R[13+ioffset,j+joffset]);
      totalSAE = totalSAE + abs(C[14,j] - R[14+ioffset,j+joffset]);
      totalSAE = totalSAE + abs(C[15,j] - R[15+ioffset,j+joffset]);
}
```

**Figure 6.21**   Pseudocode for two versions of SAE calculation

the inner loop and repeats the calculation 16 times. More lines of code are required but, on most platforms, version (b) will run faster (note that some compilers automatically unroll repetitive loops, but better performance can often be achieved by explicitly unrolling loops).

2. *'Hand-coding'* of critical operations. The SAE calculation for a block (Equation 6.3) is carried out many times during motion estimation and is therefore a candidate for coding in assembly language.

3. *Reuse of calculated values.* Consider the final stage of the TSS algorithm shown in Figure 6.9: a total of nine SAE matches are compared, each 1 pixel apart. This means that most of the operations of each SAE match are identical for each search location. It may therefore be possible to reduce the number of operations by reusing some of the calculated values $|C_{ij} - R_{ij}|$ between successive SAE calculations. (However, this may not be possible if multiple-sample calculations are used, see below.)

4. *Calculate multiple sample comparisons in a single operation.* Matching is typically carried out on 8-bit luminance samples from the current and reference frames. A single match operation $|C_{ij} - R_{ij}|$ takes as its input two 8-bit values and produces an 8-bit output value. With a large word width (e.g. 32 or 64 bits) it may be possible to carry out several matching operations at once by packing several input samples into a word.

   Figure 6.22 shows the general idea: here, four luminance samples are packed into each of two input words and the result of $|C_{ij} - R_{ij}|$ for each sample are available as the 4 bytes of an output word. Care is required with this approach: first, there is an overhead associated with packing and unpacking bytes into/out of words, and second, there may be the possibility for overflow during the comparison (since the result of $C_{ij} - R_{ij}$ is actually a 9-bit signed number prior to the magnitude operator $||$).

These and further optimisations may be applied to significantly increase the speed of the search calculation. In general, more optimisation leads to more lines of code that may be difficult to maintain and may only perform well on a particular processor platform. However, increased motion estimation performance can outweigh these disadvantages.

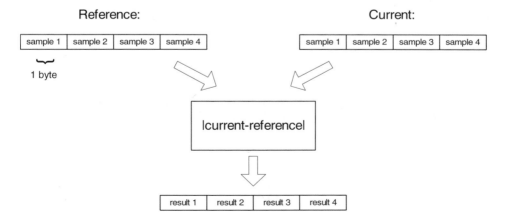

**Figure 6.22**  Multiple SAE comparisons in parallel

*Reduced complexity matching criteria*

The fast search algorithms described in Section 6.4 reduce the complexity of motion estimation by attempting to subsample the number of points in the SAE map that require to be tested. At each comparison point, Equation 6.3 must be evaluated and this requires $N \times N$ calculations (where $N$ is the block size). However, it is possible to reduce the number of calculations for each matching point in several ways.

**Early termination**  In many cases, the outcome of the SAE calculation will be an SAE that is larger than the previous minimum SAE. If we know that the current matching position will not produce the smallest SAE, we do not need to finish the calculation. If this is the case, the value total SAE in Figure 6.21 will exceed the previous minimum SAE at some point before the end of the calculation. A simple way of reducing complexity is to check for this, e.g.:

```
if (totalSAE > minSAE) break from the loop.
```

This check itself takes processing time and so it is not efficient to test after every single sample comparison: instead, a good approach is to include the above check after each inner loop (i.e. each row of 16 comparisons).

**Row and column projections**  A projection of each row and column in the current and reference blocks is formed. The projection is formed by adding all the luminance values in the current row or column: for a $16 \times 16$ block, there are 16 row projections and 16 column projections. Figure 6.23 shows the projections for one macroblock. An approximation to SAE is calculated as follows:

$$\text{SAE}_{\text{approx}} = \sum_{i=0}^{N-1} Ccol_i - Rcol_i + \sum_{j=0}^{N-1} Crow_j - Rrow_j$$

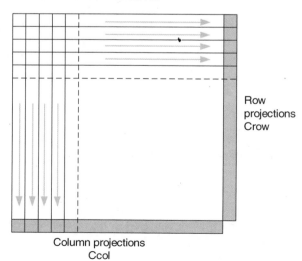

Current macroblock

Row projections Crow

Column projections Ccol

**Figure 6.23**  Row and column projections

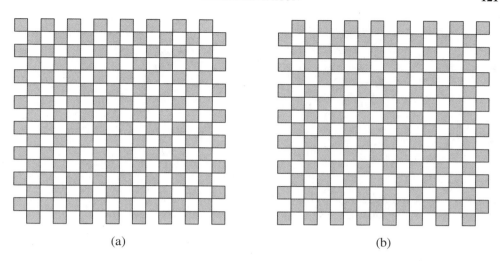

(a)                                                              (b)

**Figure 6.24**   SAE subsampling structures

where *Ccol*, *Rcol*, *Crow* and *Rrow* are the projections along the rows and columns of the current (C) and reference (R) blocks. The actual SAE calculation requires a total of 32 operations instead of 256: however, the projections themselves must also be calculated and the end result will not be as accurate as a 'true' SAE calculation.

**Subsampled SAE calculation**   Instead of calculating $|C_{ij} - R_{ij}|$ for every sample position $i, j$, an approximate SAE value may be calculated by comparing a subset of the positions $i, j$, i.e. by subsampling the current and reference blocks before carrying out the comparison. A suitable subsampling structure may be a 'chessboard' as shown in Figure 6.24. This method halves the number of calculations required. However, there are two main disadvantages. First, the method is less accurate than the full calculation. Some texture patterns in the image may cause this method to perform significantly worse than the full calculation (the performance loss may be reduced by varying the subsampling structure, for example, between the two structures shown in Figure 6.24). Second, it may be inefficient for the processor to fetch the comparison data from non-adjacent memory locations. An adaptive subsampling method has been described elsewhere.[7]

**Estimating half-pixel results**   The computational overhead of sub-pixel estimation may be reduced by interpolating between the SAE values calculated at integer positions (rather than carrying out searches at sub-pixel positions). It is clear from Figure 6.17 that the SAE is a continuous function in the region of the minimum. Figure 6.25 shows the general approach, in one dimension only. The integer SAE calculations are carried out to produce the points shown in black. By fitting a curve to the integer SAE results, estimated SAE results are found for the neighbouring half-pixel positions to the best integer result (+0.5 and +1.5). The complexity of half-pixel estimation may be significantly reduced using this method.[8,9] However, the accuracy of motion compensation depends on how accurately the curve fitting approximates the actual half-pixel results.

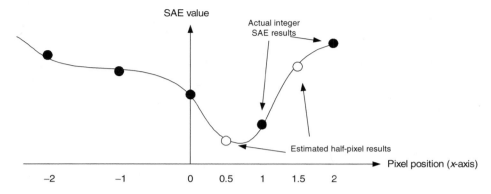

**Figure 6.25**  SAE interpolation to estimate half-pixel results

## 6.9.2  Hardware Implementations

The design of a motion estimation unit in hardware is subject to a number of (potentially conflicting) aims:

1. Maximise compression performance. A full search algorithm usually achieves the best block matching performance.

2. Minimise cycle count (and hence maximise throughput).

3. Minimise gate count.

4. Minimise data flow to/from the motion estimator.

*Example: Full search block matching unit*

A 'direct' implementation of the full search algorithm involves evaluating Equation 6.3 (SAE calculation) at each position in the search region. There are several ways in which the implementation can be speeded up (typically at the expense of matching efficiency and/or size of the design), including parallelisation (calculating multiple results in parallel), pipelining and the use of fast search algorithms.

*Parallelization of full search*

The full search algorithm is highly regular and repetitive and there are no interdependencies between the search results (i.e. the order of searches does not affect the final result). It is therefore a good candidate for parallelisation and a number of alternative approaches are available. Two popular approaches are as follows:

1. *Calculate search results in parallel.* Figure 6.26 shows the general idea: *M* processors are used, each of which calculates a single SAE result. The smallest SAE of the *M* results is

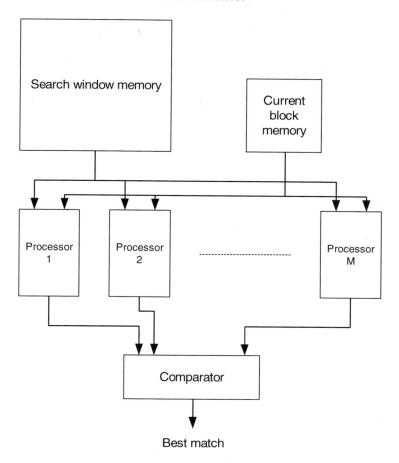

**Figure 6.26**   Parallel calculation of search results

chosen as the best match (for that particular set of calculations). The number of cycles is reduced (and the gate count of the design is increased) by a factor of approximately $M$.

2. *Calculate partial SAE results for each pixel position in parallel.* For example, the SAE calculation for a $16 \times 16$ block may be speeded up by using 16 processors, each of which calculates the SAE component for one column of pixels in the current block. Again, this approach has the potential to speed up the calculation by approximately $M$ times (if $M$ parallel processors are used).

### Fast search

It may not be feasible or practical to carry out a complete full search because of gate count or clock speed limitations. Fast search algorithms can perform almost as well as full search

**Table 6.4**  Pipelined operation: three step search

| Step 1 | Step 2 | Step 3 |
|--------|--------|--------|
| Block 1 |         |         |
| Block 2 | Block 1 |         |
| Block 3 | Block 2 | Block 1 |
| Block 4 | Block 3 | Block 2 |
| ...     | ...     | ...     |

with many fewer comparison operations and so these are attractive for hardware as well as software implementations.

In a dedicated hardware design it may be necessary to carry out each motion estimation search in a fixed number of cycles (in order to ensure that all the processing units within the design are fully utilised during encoding). In this case algorithms such as logarithmic search and nearest neighbours search are not ideal because the total number of comparisons varies from block to block. Algorithms such as the three-step search and hierarchical search are more useful because the number of operations is constant for every block.

Parallel computation may be employed to speed up the algorithm further, for example:

1. Each SAE calculation may be speeded up by using parallel processing units (each calculating the SAE for one or more columns of pixels).

2. The comparisons at one 'step' or level of the algorithm may be computed in parallel (for example, one 'step' of the three-step search or one level of the hierarchical search).

3. Successive steps of the algorithm may be pipelined to increase throughput. Table 6.4 shows an example for the three-step search. The first nine comparisons (step 1) are calculated for block 1. The next eight comparisons (step 2) for block 1 are calculated by another processing unit (or set of units), whilst step 1 is calculated for block 2, and so on.

Note that the steps or levels cannot be calculated in parallel: the search locations examined in step 2 depend on the result of step 1 and so cannot be calculated until the outcome of step 1 is known.

Option 3 above (pipelining of successive steps) is useful for sub-pixel motion estimation. Sub-pixel estimation is usually carried out on the sub-pixel positions around the best integer pixel match and this estimation step may also be pipelined. Figure 6.27 shows an example for a three-step search ($+/-7$ pixels) followed by a half-pixel estimation step. Note that *memory bandwidth* may an important issue with this type of design. Each step requires access to the current block and reference area and this can lead to an unacceptably high level of memory accesses. One option is to copy the current and reference areas to separate local memories for each processing stage but this requires more local memory.

Descriptions of hardware implementations of motion estimation algorithms can be found elsewhere.[10–12]

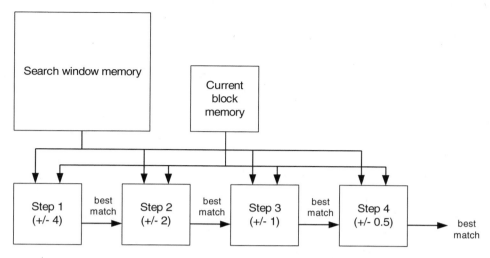

**Figure 6.27** Pipelined motion estimation: three integer steps + half-pixel step

## 6.10 SUMMARY

Motion estimation is used in an inter-frame video encoder to create a 'model' that matches the current frame as closely as possible, based on one or more previously transmitted frames ('reference frames'). This model is subtracted from the current frame (motion compensation) to produce a motion-compensated residual frame. The decoder recreates the model (based on information sent by the encoder) and adds the residual frame to reconstruct a copy of the original frame.

The goal of motion estimation design is to minimise the amount of coded information (residual frame and model information), whilst keeping the computational complexity of motion estimation and compensation to an acceptable limit. Many reduced-complexity motion estimation methods exist ('fast search' algorithms), and these allow the designer to 'trade' increased computational efficiency against reduced compression performance.

After motion estimation and compensation, the next problem faced by a video CODEC is to efficiently compress the residual frame. The most popular method is *transform coding* and this is discussed in the next chapter.

## REFERENCES

1. T. Koga, K. Iinuma et al., 'Motion compensated interframe coding for video conference', *Proc. NTC*, November 1991.
2. J. R. Jain and A. K. Jain, 'Displacement measurement and its application in interframe image coding', *IEEE Trans. Communications*, **29**, December 1981.
3. M. Ghanbari, 'The cross-search algorithm for motion estimation', *IEEE Trans. Communications*, **38**, July 1990.
4. M. Gallant, G. Côté and F. Kossentini, 'An efficient computation-constrained block-based motion estimation algorithm for low bit rate video coding', *IEEE Trans. Image Processing*, **8**(12), December 1999.

5. T. Wiegand, X. Zhang and B. Girod, 'Long-term memory motion compensated prediction', *IEEE Trans. CSVT*, September 1998.
6. B. Tao and M. Orchard, 'Removal of motion uncertainty and quantization noise in motion compensation', *IEEE Trans. CVST*, **11**(1), January 2001.
7. Y. Wang, Y. Wang and H. Kuroda, 'A globally adaptive pixel-decimation algorithm for block motion estimation', *IEEE Trans. CSVT*, **10**(6), September 2000.
8. X. Li and C. Gonzales, 'A locally quadratic model of the motion estimation error criterion function and its application to subpixel interpolation', *IEEE Trans. CSVT*, **3**, February 1993.
9. Y. Senda, 'Approximate criteria for the MPEG-2 motion estimation', *IEEE Trans. CSVT*, **10**(3), April 2000.
10. P. Pirsch, N. Demassieux and W. Gehrke, 'VLSI architectures for video compression–a survey', *Proceedings of the IEEE*, **83**(2), February 1995.
11. C. J. Kuo, C. H. Yeh and S. F. Odeh, 'Polynomial search algorithm for motion estimation', *IEEE Trans. CSVT*, **10**(5), August 2000.
12. G. Fujita, T. Onoye and I. Shirakawa, 'A VLSI architecture for motion estimation core dedicated to H.263 video coding', *IEICE Trans. Electronics*, **E81-C**(5), May 1998.

# 7

# Transform Coding

## 7.1 INTRODUCTION

Transform coding is at the heart of the majority of video coding systems and standards. Spatial image data (image samples or motion-compensated residual samples) are transformed into a different representation, the transform domain. There are good reasons for transforming image data in this way. Spatial image data is inherently 'difficult' to compress: neighbouring samples are highly correlated (interrelated) and the energy tends to be evenly distributed across an image, making it difficult to discard data or reduce the precision of data without adversely affecting image quality. With a suitable choice of transform, the data is 'easier' to compress in the transform domain. There are several desirable properties of a transform for compression. It should compact the energy in the image (concentrate the energy into a small number of significant values); it should decorrelate the data (so that discarding 'insignificant' data has a minimal effect on image quality); and it should be suitable for practical implementation in software and hardware.

The two most widely used image compression transforms are the discrete cosine transform (DCT) and the discrete wavelet transform (DWT). The DCT is usually applied to small, regular blocks of image samples (e.g. $8 \times 8$ squares) and the DWT is usually applied to larger image sections ('tiles') or to complete images. Many alternatives have been proposed, for example 3-D transforms (dealing with spatial and temporal correlation), variable block-size transforms, fractal transforms, Gabor analysis. The DCT has proved particularly durable and is at the core of most of the current generation of image and video coding standards, including JPEG, H.261, H.263, H.263+, MPEG-1, MPEG-2 and MPEG-4. The DWT is gaining popularity because it can outperform the DCT for still image coding and so it is used in the new JPEG image coding standard (JPEG-2000) and for still 'texture' coding in MPEG-4.

This chapter concentrates on the DCT. The theory and properties of the transforms are described first, followed by an introduction to practical algorithms and architectures for the DCT. Closely linked with the DCT is the process of quantisation and the chapter ends with a discussion of quantisation theory and practice.

## 7.2 DISCRETE COSINE TRANSFORM

Ahmed, Natarajan and Rao originally proposed the DCT in 1974.[1] Since then, it has become the most popular transform for image and video coding. There are two main reasons for its popularity: first, it is effective at transforming image data into a form that is easy to compress and second, it can be efficiently implemented in software and hardware.

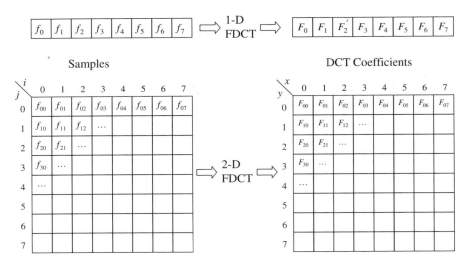

**Figure 7.1** 1-D and 2-D discrete cosine transform

The forward DCT (FDCT) transforms a set of image samples (the 'spatial domain') into a set of transform coefficients (the 'transform domain'). The transform is reversible: the inverse DCT (IDCT) transforms a set of coefficients into a set of image samples. The forward and inverse transforms are commonly used in 1-D or 2-D forms for image and video compression. The 1-D version transforms a 1-D array of samples into an a 1-D array of coefficients, whereas the 2-D version transforms a 2-D array (block) of samples into a block of coefficients. Figure 7.1 shows the two forms of the DCT.

The DCT has two useful properties for image and video compression, *energy compaction* (concentrating the image energy into a small number of coefficients) and *decorrelation* (minimising the interdependencies between coefficients). Figure 7.2 illustrates the energy compaction property of the DCT. Image (a) is an $80 \times 80$ pixel image and image (b) plots the coefficients of the 2-D DCT. The energy in the transformed coefficients is concentrated about the top-left corner of the array of coefficients (*compaction*). The top-left coefficients correspond to low frequencies: there is a 'peak' in energy in this area and the coefficient values rapidly decrease to the bottom right of the array (the higher-frequency coefficients). The DCT coefficients are *decorrelated* which means that many of the coefficients with small values can be discarded without significantly affecting image quality. A compact array of decorrelated coefficients can be compressed much more efficiently than an array of highly correlated image pixels.

The decorrelation and compaction performance of the DCT increases with block size. However, computational complexity also increases (exponentially) with block size. A block size of $8 \times 8$ is commonly used in image and video coding applications. This size gives a good compromise between compression efficiency and computational efficiency (particularly as there are a number of efficient algorithms for a DCT of size $2^m \times 2^m$, where $m$ is an integer).

The forward DCT for an $8 \times 8$ block of image samples is given by Equation 7.1:

$$F_{x,y} = \frac{C(x)C(y)}{4} \sum_{i=0}^{7} \sum_{j=0}^{7} f_{i,j} \cos\left(\frac{(2i+1)x\pi}{16}\right) \cos\left(\frac{(2j+1)y\pi}{16}\right) \qquad (7.1)$$

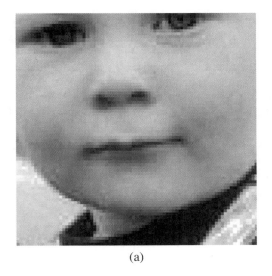

(a)

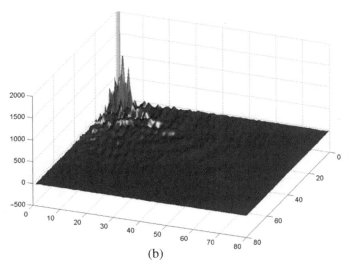

(b)

**Figure 7.2** (a) 80 × 80 pixel image; (b) 2-D DCT

$f_{i,j}$ are the 64 samples $(i,j)$ of the input sample block, $F_{x,y}$ are the 64 DCT coefficients $(x,y)$ and $C(x)$, $C(y)$ are constants:

$$C(n) = \begin{cases} \dfrac{1}{\sqrt{2}}, & n = 0 \\ 1, & n \neq 0 \end{cases}$$

The labelling of samples $(f_{i,j})$ and coefficients $(F_{x,y})$ is illustrated in Figure 7.1.

The DCT represents each block of image samples as a weighted sum of 2-D cosine functions ('basis functions'). The functions are plotted as surfaces in Figure 7.3 and as $8 \times 8$ pixel 'basis patterns' in Figure 7.4. The top-left pattern has the lowest 'frequency' and is just a uniform block. Moving to the right, the patterns contain an increasing number of 'cycles' between dark and light in the horizontal direction: these represent increasing horizontal spatial frequency. Moving down, the patterns contain increasing vertical spatial frequency. Moving diagonally to the right and down, the patterns contain both horizontal and vertical frequencies. The block of samples may be reconstructed by adding together the 64 basis patterns, each multiplied by a weight (the corresponding DCT coefficient $F_{x,y}$).

The inverse DCT reconstructs a block of image samples from an array of DCT coefficients. The IDCT takes as its input a block of $8 \times 8$ DCT coefficients $F_{x,y}$ and reconstructs a block of $8 \times 8$ image samples $f_{i,j}$ (Equation 7.2).

$$f_{i,j} = \sum_{x=0}^{7} \sum_{y=0}^{7} \frac{C(x)C(y)}{4} F_{x,y} \cos\left(\frac{(2i+1)x\pi}{16}\right) \cos\left(\frac{(2j+1)y\pi}{16}\right) \quad (7.2)$$

$C(x)$ and $C(y)$ are the same constants as for the FDCT.

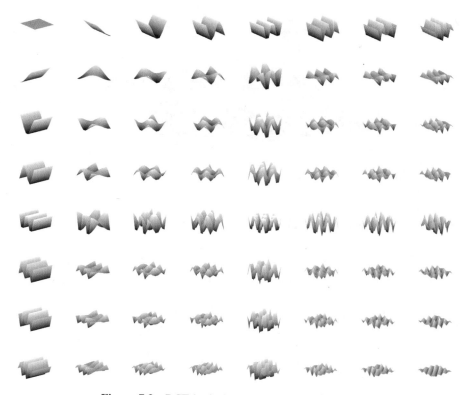

**Figure 7.3**  DCT basis functions (plotted as surfaces)

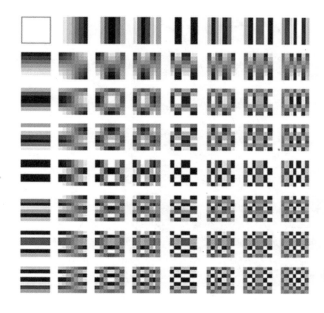

**Figure 7.4**   DCT basis patterns

## *Example: DCT of an 8 × 8 block of samples*

Figure 7.5 shows an 8 × 8 block of samples (b) taken from image (a). The block is transformed with a 2-D DCT to produce the coefficients shown in image (c). The six most significant coefficients are: (0, 0), (1, 0), (1, 1), (2, 0), (3, 0) and (4, 0) and these are highlighted on the table of coefficients (Table 7.1).

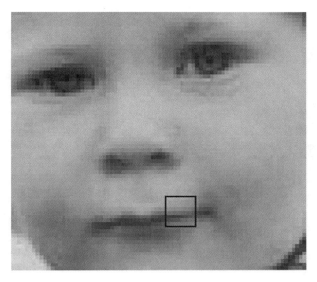

(a)

**Figure 7.5**   (a) Original image; (b) 8 × 8 block of samples; (c) 2-D DCT coefficients

Original block

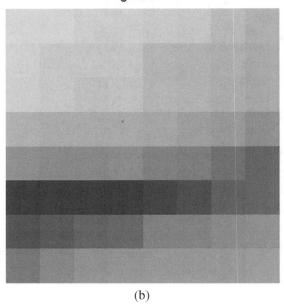

(b)

DCT coefficients

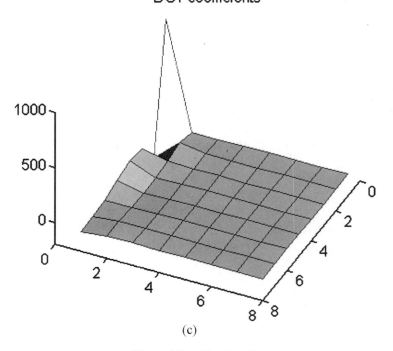

(c)

**Figure 7.5**   (*Continued*)

**Table 7.1** DCT coefficients

| | | | | | | | |
|---|---|---|---|---|---|---|---|
| 967.5 | − 6.3 | − 10.7 | − 2.7 | − 1.2 | − 1.1 | − 1.3 | 0.1 |
| − 163.4 | − 71.3 | 1.8 | 4.2 | − 1.8 | 2.9 | 0.2 | − 1.6 |
| 55.3 | 13.6 | − 1.2 | 2.3 | − 0.6 | 2.0 | 0.7 | − 0.6 |
| 81.8 | − 8.9 | − 4.7 | 1.4 | 0.4 | − 0.5 | − 0.1 | 1.0 |
| 38.9 | − 12.9 | − 7.0 | − 1.3 | − 0.6 | 0.1 | − 0.3 | 1.0 |
| − 7.6 | − 8.4 | 0.8 | 2.6 | 0.6 | − 1.9 | 0.1 | 2.1 |
| − 14.7 | 1.4 | 4.1 | 1.2 | 0.0 | − 0.3 | − 0.2 | 1.1 |
| − 22.3 | 5.0 | 4.3 | 1.3 | − 0.6 | − 0.5 | 0.2 | − 0.3 |

A reasonable approximation to the original image block can be reconstructed from just these six coefficients, as shown in Figure 7.6. First, coefficient (0, 0) is multiplied by a weight of 967.5 and transformed with the inverse DCT. This coefficient represents the average 'shade' of the block (in this case, mid-grey) and is often described as the 'DC coefficient' (the DC coefficient is usually the most significant in any block). Figure 7.6 shows the reconstructed block formed by the DC coefficient only (the top-right block in the figure). Next, coefficient (1, 0) is multiplied by a weight of − 163.4 (equivalent to *subtracting* its basis pattern). The weighted basis pattern is shown in the second row of Figure 7.6 (on the left) and the *sum* of the first two patterns is shown on the right. As each of the further four basis patterns is added to the reconstruction, more detail is added to the reconstructed block. The final result (shown on the bottom right of Figure 7.6 and produced using just 6 out of the 64 coefficients) is a good approximation of the original. This example illustrates the two key properties of the DCT: the significant coefficients are clustered around the DC coefficient (compaction) and the block may be reconstructed using only a small number of coefficients (decorrelation).

## 7.3 DISCRETE WAVELET TRANSFORM

The DCT described above operates on a block of samples (usually $16 \times 16$, $8 \times 8$ or smaller) and decomposes these samples into a set of spatial frequency components. A wavelet transform also decomposes spatial image data according to frequency and wavelet-based compression has been shown to outperform DCT-based compression for still images. Because of this, the new version of the JPEG image compression standard, JPEG-2000, is wavelet-based rather than DCT-based.

One of the key differences between the application of wavelet and discrete cosine transforms is that a wavelet transform is typically applied to a complete image or a large

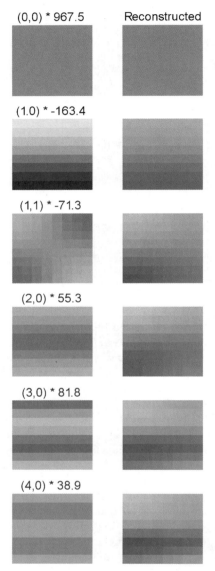

**Figure 7.6** Reconstruction of image block from six basis patterns

rectangular region ('tile') of the image, in contrast with the small block size chosen for DCT implementations. The DCT becomes increasingly complex to calculate for larger block sizes, whilst the benefits of larger blocks rapidly diminish above $8 \times 8$ or $16 \times 16$ samples, whereas a wavelet transform may be more efficiently applied to large blocks or complete images and produces better compression performance for larger blocks.

A single-stage wavelet transformation consists of a filtering operation that decomposes an image into four frequency bands as shown in Figure 7.7. Image (a) is the original; image (b) is the result of a single-stage wavelet transform. The top-left corner of the transformed image ('LL') is the original image, low-pass filtered and subsampled in the horizontal and vertical

(a)

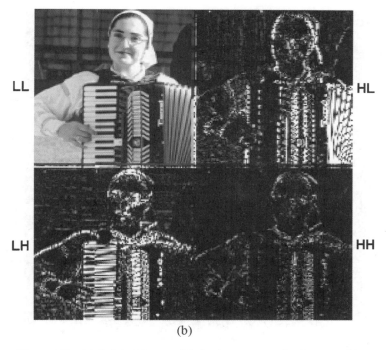

(b)

**Figure 7.7** (a) Original image; (b) single-stage wavelet decomposition

dimensions. The top-right corner ('HL') consists of residual vertical frequencies (i.e. the vertical component of the difference between the subsampled 'LL' image and the original image). The bottom-left corner 'LH' contains residual horizontal frequencies (for example, the accordion keys are very visible here), whilst the bottom-right corner 'HH' contains residual diagonal frequencies.

This decomposition process may be repeated for the 'LL' component to produce another set of four components: a new 'LL' component that is a further subsampled version of the original image, plus three more residual frequency components. Repeating the decomposition three times gives the wavelet representation shown in Figure 7.8. The small image in the top left is the low-pass filtered original and the remaining squares contain progressively higher-frequency residual components. This process may be repeated further if desired (until, in the limit, the top-left component contains only 1 pixel which is equivalent to the 'DC' or average value of the entire image). Each sample in Figure 7.8 represents a wavelet transform *coefficient*.

The wavelet decomposition has some important properties. First, the number of wavelet 'coefficients' (the spatial values that make up Figure 7.8) is the same as the number of pixels in the original image and so the transform is not inherently adding or removing information. Second, many of the coefficients of the high-frequency components ('HH', 'HL' and 'LH' at each stage) are zero or insignificant. This reflects the fact that much of the important information in an image is low-frequency. Our response to an image is based upon a low-frequency 'overview' of the image with important detail added by higher frequencies in a few significant areas of the image. This implies that it should be possible to efficiently

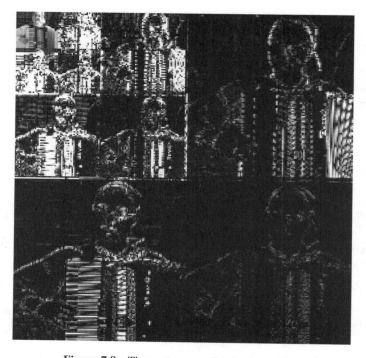

**Figure 7.8**   Three-stage wavelet decomposition

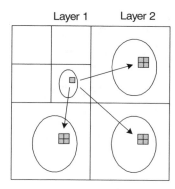

**Figure 7.9** Relationship between 'parent' and 'child' regions

compress the wavelet representation shown in Figure 7.8 if we can discard the insignificant higher-frequency coefficients whilst preserving the significant ones. Third, the decomposition is not restricted by block boundaries (unlike the DCT) and hence may be a more flexible way of decorrelating the image data (i.e. concentrating the significant components into a few coefficients) than the block-based DCT.

The method of representing significant coefficients whilst discarding insignificant coefficients is critical to the use of wavelets in image compression. The embedded zero tree approach and, more recently, set partitioning into hierarchical trees (SPIHT) are considered by some researchers to be the most effective way of doing this.[2] The wavelet decomposition can be thought of as a 'tree', where the 'root' is the top-left LL component and its 'branches' are the successively higher-frequency LH, HL and HH components at each layer. Each coefficient in a low-frequency component has a number of corresponding 'child' coefficients in higher-frequency components. This concept is illustrated in Figure 7.9, where a single coefficient at layer 1 maps to four 'child' coefficients in each component at layer 2. Zero-tree coding works on the principle that if a parent coefficient is visually insignificant then its 'children' are unlikely to be significant. Working from the top left, each coefficient and its children are encoded as a 'tree'. As soon as the tree reaches a coefficient that is insignificant, that coefficient and all its children are coded as a 'zero tree'. The decoder will reconstruct the significant coefficients and set all coefficients in a 'zero tree' to zero.

This approach provides a flexible and powerful method of image compression. The decision as to whether a coefficient is 'significant' or 'insignificant' is made by comparing it with a threshold. Setting a high threshold means that most of the coefficients are discarded and the image is highly compressed; setting a low threshold means that most coefficients are retained, giving low compression and high image fidelity. This process is equivalent to quantisation of the wavelet coefficients.

Wavelet-based compression performs well for still images (particularly in comparison with DCT-based compression) and can be implemented reasonably efficiently. Under high compression, wavelet-compressed images do not exhibit the blocking effects characteristic of the DCT. Instead, degradation is more 'graceful' and leads to a gradual blurring of the image as higher-frequency coefficients are discarded. Figure 7.10 compares the results of compression of the original image (on the left) with a DCT-based algorithm (middle image, JPEG compression) and a wavelet-based algorithm (right-hand image, JPEG-2000 compression). In

|(a)|(b)|(c)|

**Figure 7.10** (a) Original; (b) compressed and decompressed (DCT); (c) compressed and decompressed (wavelet)

each case, the compression ratio is 16×. The decompressed JPEG image is clearly distorted and 'blocky', whereas the decompressed JPEG-2000 image is much closer to the original.

Because of its good performance in compressing images, the DWT is used in the new JPEG-2000 still image compression standard and is incorporated as a still image compression tool in MPEG-4 (see Chapter 4). However, wavelet techniques have not yet gained widespread support for motion video compression because there is not an easy way to extend wavelet compression in the temporal domain. Block-based transforms such as the DCT work well with block-based motion estimation and compensation, whereas efficient, computationally tractable motion-compensation methods suitable for wavelet-based compression have not yet been demonstrated. Hence, the DCT is still the most popular transform for video coding applications.

## 7.4   FAST ALGORITHMS FOR THE DCT

According to Equation 7.1, each of the 64 coefficients in the FDCT is a weighted function of all 64 image samples. This means that 64 calculations, each involving a multiplication and an accumulation ('multiply–accumulate') must be carried out for each DCT coefficient. A total of $64 \times 64 = 4096$ multiply–accumulate operations are required for a full $8 \times 8$ FDCT. Fortunately, the DCT lends itself to significant simplification.

### 7.4.1   Separable Transforms

Many practical implementations of the FDCT and IDCT use the separable property of the transforms to simplify calculation. The 2-D FDCT can be calculated by repeatedly applying the 1-D DCT. The 1-D FDCT is given by Equation 7.3:

$$F_x = \frac{C(x)}{2} \sum_{i=0}^{7} f_i \cos\left(\frac{(2i+1)x\pi}{16}\right) \tag{7.3}$$

where $F_x (x = 0$ to 7) are the eight coefficients, $f_i$ are the eight input samples and $C(x)$ is a constant as before.

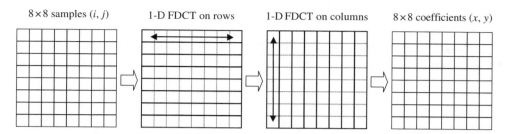

8×8 samples $(i, j)$     1-D FDCT on rows     1-D FDCT on columns     8×8 coefficients $(x, y)$

**Figure 7.11**   2-D DCT via two 1-D transforms

Equation 7.1 may be rearranged as follows (Equation 7.4):

$$F_{x,y} = \frac{C(y)}{2} \sum_{j=0}^{7} F_x \cos \left( \frac{(2j+1)y\pi}{16} \right) \tag{7.4}$$

where $F_x$ is the 1-D DCT described by Equation 7.3. In other words, the 2-D DCT can be represented as:

$$F_{x,y} = \text{1-D DCT}_{y\text{-direction}} \left( \text{1-D DCT}_{x\text{-direction}} \right)$$

The 2-D DCT of an $8 \times 8$ block can be calculated in two passes: a 1-D DCT of each row, followed by a 1-D DCT of each column (or vice versa). This property is known as *separability* and is shown graphically in Figure 7.11.

The 2-D IDCT can be calculated using two 1-D IDCTs in a similar way. The equation for the 1-D IDCT is given by Equation 7.5:

$$f_i = \sum_{x=0}^{7} \frac{C(x)}{2} F_x \cos \left( \frac{(2i+1)x\pi}{16} \right) \tag{7.5}$$

This separable approach has two advantages. First, the number of operations is reduced: each 'pass' requires $8 \times 64$ multiply–accumulate operations, i.e. the total number of operations is $2 \times 8 \times 64 = 1024$. Second, the 1-D DCT can be readily manipulated to streamline the number of operations further or to otherwise simplify calculation.

Practical implementations of the FDCT and IDCT fall into two main categories:

1. Minimal computation: the DCT operations (1-D or 2-D) are reorganised to exploit the inherent symmetry of cosine operations in order to minimise the number of multiplications and/or additions. This approach is appropriate for software implementations.

2. Maximal regularity: the 1-D or 2-D calculations are organised to regularise the data flow and processing order. This approach is suitable for dedicated hardware implementations.

In general, 1-D implementations (using the separable property described above) are less complex than 2-D implementations, but it is possible to achieve higher performance by manipulating the 2-D equations directly.

## 7.4.2  Flowgraph Algorithms

The computational requirements of the 1-D DCT may be reduced significantly by exploiting the symmetry of the cosine weighting factors. We show how the complexity of the DCT can be reduced in the following example.

*Example: Calculating 1-D DCT coefficient $F_2$*

From Equation 7.3:

$$F_2 = \frac{1}{2}\sum_{i=0}^{7} f_i \cos\left(\frac{(2i+1).2\pi}{16}\right) \tag{7.6}$$

Expanding Equation 7.6 gives

$$F_2 = \frac{1}{2}\left[ f_0 \cos\left(\frac{\pi}{8}\right) + f_1 \cos\left(\frac{3\pi}{8}\right) + f_2 \cos\left(\frac{5\pi}{8}\right) + f_3 \cos\left(\frac{7\pi}{8}\right) + f_4 \cos\left(\frac{9\pi}{8}\right) \right.$$
$$\left. + f_5 \cos\left(\frac{11\pi}{8}\right) + f_6 \cos\left(\frac{13\pi}{8}\right) + f_7 \cos\left(\frac{15\pi}{8}\right) \right] \tag{7.7}$$

The following properties of the cosine function can be used to simplify Equation 7.7:

$$\cos\left(\frac{15\pi}{8}\right) = -\cos\left(\frac{9\pi}{8}\right) = -\cos\left(\frac{7\pi}{8}\right) = \cos\left(\frac{\pi}{8}\right)$$

and

$$\cos\left(\frac{13\pi}{8}\right) = -\cos\left(\frac{11\pi}{8}\right) = -\cos\left(\frac{5\pi}{8}\right) = \cos\left(\frac{3\pi}{8}\right)$$

These relationships are shown graphically in Figure 7.12 where $\cos(\pi/8)$, $\cos(7\pi/8)$, etc. are plotted as circles ($\circ$) and $\cos(3\pi/8)$, $\cos(5\pi/8)$, etc. are plotted as stars ($\star$).
Using the above relationships, Equation 7.7 can be rewritten as follows:

$$F_2 = \frac{1}{2}\left[ f_0 \cos\left(\frac{\pi}{8}\right) + f_1 \cos\left(\frac{3\pi}{8}\right) - f_2 \cos\left(\frac{3\pi}{8}\right) - f_3 \cos\left(\frac{\pi}{8}\right) - f_4 \cos\left(\frac{\pi}{8}\right) \right.$$
$$\left. - f_5 \cos\left(\frac{3\pi}{8}\right) + f_6 \cos\left(\frac{3\pi}{8}\right) + f_7 \cos\left(\frac{\pi}{8}\right) \right] \tag{7.8}$$

hence:

$$2F_2 = \left[ (f_0 - f_4 + f_7 - f_3) \cdot \cos\left(\frac{\pi}{8}\right) + (f_1 - f_2 - f_5 + f_6) \cdot \cos\left(\frac{3\pi}{8}\right) \right] \tag{7.9}$$

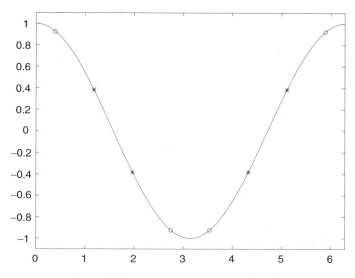

**Figure 7.12**  Symmetries of cosine function

The calculation for $F_2$ has been reduced from eight multiplications and eight additions (Equation 7.7) to two multiplications and eight additions/subtractions (Equation 7.9).

Applying a similar process to $F_6$ gives

$$2F_6 = \left[ (f_0 - f_4 + f_7 - f_3) \cdot \cos\left(\frac{3\pi}{8}\right) + (f_1 - f_2 - f_5 + f_6) \cdot \cos\left(\frac{\pi}{8}\right) \right] \qquad (7.10)$$

The additions and subtractions are clearly the same as in Equation 7.9. We can therefore combine the calculations of $F_2$ and $F_6$ as follows:

$$\text{Step 1}: \text{ calculate } \quad D_1 = (f_0 - f_4 + f_7 - f_3) \quad \text{and} \quad D_2 = (f_1 - f_2 + f_5 + f_6)$$

$$\text{Step 2}: \text{ calculate } 2F_2 = \left[ D_1 \cos\left(\frac{\pi}{8}\right) + D_2 \cos\left(\frac{3\pi}{8}\right) \right] \quad \text{and}$$

$$2F_6 = \left[ D_1 \cos\left(\frac{3\pi}{8}\right) + D_2 \cos\left(\frac{\pi}{8}\right) \right]$$

In total, the two steps require 8 additions or subtractions and 4 multiplications, compared with 16 additions and 16 multiplications for the full calculation. The combined calculations of $F_2$ and $F_6$ can be graphically represented by a *flowgraph* as shown in Figure 7.13. In this figure, a circle represents addition and a square represents multiplication by a scaling factor. For clarity, the cosine scaling factors are represented as 'cX', meaning 'multiply by $\cos(X\pi/16)$'. Hence, $\cos(\pi/8)$ is represented by c2 and $\cos(3\pi/8)$ is represented by c6.

This approach can be extended to simplify the calculation of $F_0$ and $F_4$, producing the *top* half of the flowgraph shown in Figure 7.14. Applying basic cosine symmetries does not give

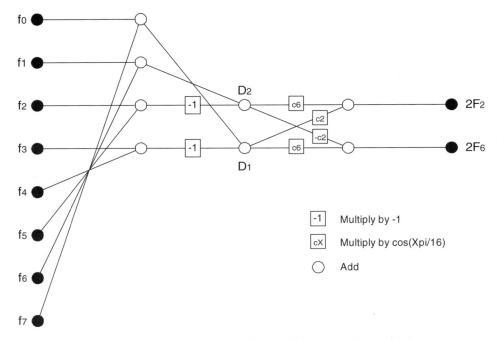

**Figure 7.13**   Partial FDCT flowgraph (F2 and F6 outputs)

such a useful result for the odd-numbered FDCT coefficients (1, 3, 5, 7). However, further manipulation of the matrix of cosine weighting factors can simplify the calculation of these coefficients. Figure 7.14 shows a widely used example of a 'fast' DCT algorithm,[3] requiring only 26 additions/subtractions and 20 multiplications (in fact, this can be reduced to 16 multiplications by combining some of the multiplications by c4). This is considerably simpler than the 64 multiplies and 64 additions of the 'direct' 1-D DCT. Each multiplication is by a constant scaling factor and these scaling factors may be pre-calculated to speed up computation.

In Figure 7.14, eight samples $f_0 \ldots f_7$ are input at the left and eight DCT coefficients $2F_0 \ldots 2F_7$ are output at the right. A 1-D IDCT may be carried out by simply reversing the direction of the graph, i.e. the coefficients $2F_0 \ldots 2F_7$ are now inputs and the samples $f_0 \ldots f_7$ are outputs.

By manipulating the transform operations in different ways, many other flowgraph algorithms can be obtained. Each algorithm has characteristics that may make it suitable for a particular application: for example, minimal number of multiplications (for processing platforms where multiplications are particularly expensive), minimal total number of operations (where multiplications are not computationally expensive), highly regular data flow, and so on. Table 7.2 summarises the features of some popular 1-D 'fast' algorithms. Arai's algorithm requires only five multiplications, making it very efficient for most processing platforms; however, this algorithm results in incorrectly scaled coefficients and this must be compensated for by scaling the quantisation algorithm (see Section 7.6).

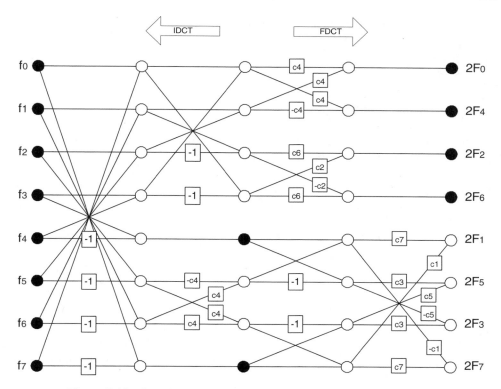

**Figure 7.14** Complete FDCT flowgraph (from Chen, Fralick and Smith)

These fast 1-D algorithms exploit symmetries of the 1-D DCT and there are further variations on the fast 1-D DCT .[7,8] In order to calculate a complete 2-D DCT, the 1-D transform is applied independently to the rows and then to the columns of the block of data. Further reductions in the number of operations may be achieved by taking advantage of further symmetries in the 'direct' form of the 2-D DCT (Equation 7.1). In general, it is possible to obtain better performance with a direct 2-D algorithm[9,10] than with separable 1-D algorithms. However, this improved performance comes at the cost of a significant increase in algorithmic complexity. In many practical applications, the relative simplicity (and smaller software code size) of the 1-D transforms is preferable to the higher complexity

**Table 7.2** Comparison of 1-D DCT algorithms (8-point DCT)

| Source | Multiplications | Additions |
|---|---|---|
| 'Direct' | 64 | 64 |
| Chen[3] | 16 | 26 |
| Lee[4] | 12 | 29 |
| Loeffler[5] | 11 | 29 |
| Arai[6] | 5 | 28 |

of direct 2-D algorithms. For example, it is more straightforward to develop a highly efficient hand-optimised 1-D function than to carry out the same software optimisations with a larger 2-D function.

### 7.4.3 Distributed Algorithms

The 'flowgraph' algorithms described above are widely used in software implementations of the DCT but have two disadvantages for dedicated hardware designs. They tend to be irregular (i.e. different operations take place at each stage of the flowgraph) and they require large multipliers (which take up large areas of silicon in an IC). It is useful to develop alternative algorithms that have a more regular structure and/or do not require large parallel multipliers.

Equation 7.3 (1-D FDCT) may be written as a 'sum of products':

$$F_x = \sum_{i=0}^{7} C_{i,x} f_i \quad \text{where} \quad C_{i,x} = \frac{C(x)}{2} \cos\left(\frac{(2i+1)x\pi}{16}\right) \tag{7.11}$$

The 1-D FDCT is the sum of eight products, where each product term is formed by multiplying an input sample by a constant weighting factor $C_{i,x}$. The first stage of Chen's fast algorithm shown in Figure 7.14 is a series of additions and subtractions and these can be used to simplify Equation 7.11. First, calculate four sums ($u$) and four differences ($v$) from the input data:

$$u_i = f_i + f_{7-i}$$
$$v_i = f_i - f_{7-i} \quad (i = 0, 1, 2, 3) \tag{7.12}$$

Equation 7.11 can be decomposed into two smaller calculations:

$$F_x = \sum_{i=0}^{3} C_{i,x} u_i \quad (x = 0, 2, 4, 6) \tag{7.13}$$

$$F_x = \sum_{i=0}^{3} C_{i,x} v_i \quad (x = 1, 3, 5, 7) \tag{7.14}$$

In this form, the calculations are suitable for implementation using a technique known as *distributed arithmetic* (first proposed in 1974 for implementing digital filters[11]). Each multiplication is carried out a bit at a time, using a look-up table and an accumulator (rather than a parallel multiplier).

A $B$-bit twos complement binary number $n$ can be represented as:

$$n = -n^0 + \sum_{j=1}^{B-1} 2^{-j} n^j \tag{7.15}$$

$n^0$ is the most significant bit (MSB) of $n$ (the sign bit) and $n^j$ are the remaining $(B-1)$ bits of $n$.

Assuming that each input $u_i$ is a $B$-bit binary number in twos complement form, Equation 7.13 becomes

$$F_x = \sum_{i=0}^{3} C_{i,x}\left(-u_i^0 + \sum_{j=1}^{B-1} 2^{-j}u_i^j\right) \qquad (7.16)$$

Rearranging gives

$$F_x = \sum_{j=1}^{B-1} 2^{-j}\left(\sum_{i=0}^{3} C_{i,x}u_i^j\right) - \left(\sum_{i=0}^{3} C_{i,x}u_i^0\right) \qquad (7.17)$$

or

$$F_x = \sum_{j=1}^{B-1} 2^{-j}D_x(u^j) - D_x(u^0) \quad \text{where } D_x(u^j) = \left(\sum_{i=0}^{3} C_{i,x}u_i^j\right) \qquad (7.18)$$

$D_x(u^j)$ is a function of the bits at position $j$ in each of the four input values: these bits are $u_0^j$, $u_1^j$, $u_2^j$ and $u_3^j$. This means that there are only $2^4 = 16$ possible outcomes of $D_x$ and these 16 outcomes may be pre-calculated and stored in a look-up table. The FDCT describe by Equation 7.18 can be carried out by a series of table look-ups ($D_x$), additions ($\sum$) and shifts ($2^{-j}$). In this form, no multiplication is required and this maps efficiently to hardware (see Section 7.5.2). A similar approach is taken to calculate the four odd-numbered FDCT coefficients $F_1$, $F_3$, $F_5$ and $F_7$ and the distributed form may also be applied to the 1-D IDCT.

## 7.4.4 Other DCT Algorithms

The popularity of the DCT has led to the development of further algorithms. For example, a 1-D DCT in the form of a finite difference equation has been presented.[12] Using this form, the DCT coefficients may be calculated recursively using an infinite impulse response (IIR) filter. This has several advantages: the algorithm is very regular and there are a number of well-established methods for implementing IIR filters in hardware and software.

Recently, approximate forms of the DCT have been proposed. Each of the DCT-based image and video coding standards specifies a minimum accuracy for the inverse DCT and in order to meet this specification it is necessary to use multiplications and fractional-precision numbers. However, if accuracy and/or complete compatibility with the standards are of lesser importance, it is possible to calculate the DCT and IDCT using one of several approximations. For example, an approximate algorithm has been proposed[13] that requires only additions and shifts (i.e. there are no multiplications). This type of approximation may be suitable for low-complexity software or hardware implementations where computational power is very limited. However, the disadvantage is that image quality will be reduced compared with an accurate DCT implementation. An interesting trend is shown in the H.26L draft standard (see Chapter 5), where an integer DCT approximation is defined as part of the standard to facilitate low-complexity implementations whilst retaining compliance with the standard.


## 7.5  IMPLEMENTING THE DCT

### 7.5.1  Software DCT

The choice of 'fast' algorithm for a software video CODEC depends on a number of factors. Different processing platforms (see Chapter 12) have different strengths and weaknesses and these may influence the choice of DCT/IDCT algorithm. Factors include:

- Computational 'cost' of multiplication. Some processors take many cycles to carry out a multiplication, others are reasonably fast. Alternative flowgraph-based algorithms allow the designer to 'trade' the number of multiplications against the total number of operations.

- Fixed vs. floating-point arithmetic capabilities. Poor floating-point performance may be compensated for by scaling the DCT multiplication factors to integer values.

- Register availability. If the processor has a small number of internal registers then temporary variables should be kept to a minimum and reused where possible.

- Availability of dedicated operations. 'Custom' operations such as digital signal processor (DSP) multiply–accumulate operations and the Intel MMX instructions may be used to improve the performance of some DCT algorithms (see Chapter 12).

Because of the proliferation of 'fast' algorithms, it is usually possible to choose a 'shortlist' of two or three alternative algorithms (typically flowgraph-based algorithms for software designs) and to compare the performance of each algorithm on the target processor before making the final choice.


*Example*

Figure 7.15 lists pseudocode for Chen's algorithm (shown in Figure 7.14). (Only the top-half calculations are given for clarity). The multiplication factors cX are pre-calculated constants. In this example, floating-point arithmetic is used: alternatively, the multipliers cX may be scaled up to integers and the entire DCT may be carried out using integer arithmetic (in which case, the final results must be scaled back down to compensate). The cosine multiplication factors never change and so these may be pre-calculated (in this case as floating-point numbers). A 1-D DCT is applied to each row in turn, then to each column. Note the use of a reasonably large number of temporary variables.

Further performance optimisation may be achieved by exploiting the flexibility of a software implementation. For example, variable-complexity algorithms (VCAs) may be applied to reduce the number of operations required to calculate the DCT and IDCT (see Chapter 10 for some examples).

```
constant c4 = 0.707107

constant c2 = 0.923880

constant c6 = 0.382683

// (similarly for c1, c3, c5 and c7)

for (every row) {

        i0 = f0+f7      // First stage

        i1 = f1+f6

        i2 = f2+f5

        i3 = f3+f4

        i4 = f3-f4

        i5 = f2-f5

        i6 = f1-f6

        i7 = f0-f7

        j0 = i0 + i3    // Second stage

        j1 = i1+ i2

        j2 = i1 - i2

        j3 = i0 - i3

        // (similarly for j4..j7)

        k0 = (j0 + j1) * c4    // Third stage

        k1 = (j0 - j1) * c4

        k2 = (j2*c6) + (j3*c2)

        k3 = (j3*c6) - (j2*c2)

        // (similarly for k4..k7)

        F0 = k0>>1

        F4 = k1>>1

        F2 = k2>>1

        F6 = k3>>1

        // (F1..F7 require another stage of multiplications and additions)

        ...

}       // end of row calculations

for (every column) {

        // repeat above steps on the columns

}
```

**Figure 7.15**   Pseudocode for Chen's algorithm

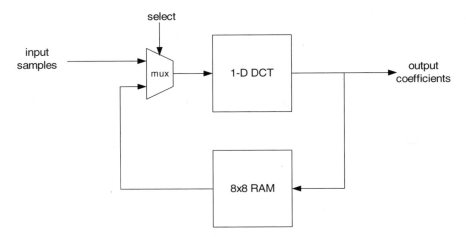

**Figure 7.16**    2-D DCT architecture

## 7.5.2   Hardware DCT

Dedicated hardware implementations of the FDCT/IDCT (suitable for ASIC or FPGA designs, for example) typically make use of separable 1-D transforms to calculate the 2-D transform. The two sets of row/column transforms shown in Figure 7.11 may be carried out using a single 1-D transform unit by transposing the $8 \times 8$ array between the two 1-D transforms, i.e.

Input data $\rightarrow$ 1-D transform on rows $\rightarrow$ Transpose array
$\rightarrow$ 1-D transform on columns $\rightarrow$ Output data

An $8 \times 8$ RAM ('transposition RAM') may be used to carry out the transposition. Figure 7.16 shows an architecture for the 2-D DCT that uses a 1-D transform 'core' together with a transposition RAM. The following stages are required to calculate a complete 2-D FDCT (or IDCT):

1. Load input data in row order; calculate 1-D DCT of each row; write into transposition RAM in row order.

2. Read from RAM in column order; calculate 1-D DCT of each column; write into RAM in column order.

3. Read output data from RAM in row order.

There are a number of options for implementing the 1-D FDCT or IDCT 'core'. Flowgraph algorithms are not ideal for hardware designs: the data flow is not completely regular and it is not usually possible to efficiently reuse arithmetic units (such as adders and multipliers). Two popular and widely used designs are the *parallel multiplier* and *distributed arithmetic* approaches.

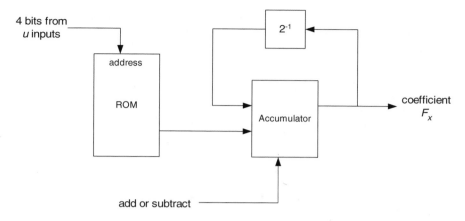

4 bits from
*u* inputs

address

ROM

$2^{-1}$

Accumulator

coefficient
$F_x$

add or subtract

**Figure 7.17**   Distributed arithmetic ROM-accumulator

## *Parallel multiplier*

This is a more or less direct implementation of Equations 7.13 and 7.14 (four-point sum of products). After an initial stage that calculates the four sums (*u*) and differences (*v*) (see Equation 7.12), each sum-of-products result is calculated. There are 16 possible factors $C_{i,x}$ for each of the two 4-point DCTs, and these factors may be pre-calculated to simplify the design. High performance may be achieved by carrying out the four multiplications for each result in parallel; however, this requires four large parallel multipliers and may be expensive in terms of logic gates.

## *Distributed arithmetic*

The basic calculation of the distributed arithmetic algorithm is given by Equation 7.18. This calculation maps to the hardware circuit shown in Figure 7.17, known as a *ROM-Accumulator* circuit. Calculating each coefficient $F_x$ takes a total of $B$ clock cycles and proceeds as follows (Table 7.3). The accumulator is reset to zero at the start. During each

**Table 7.3**   Distributed arithmetic: calculation of one coefficient

| Bit position | ROM input | ROM output | Accumulator contents |
|---|---|---|---|
| $B-1$ | $u^{B-1}$ (i.e. bit $(B-1)$ of $u_0$, $u_1$, $u_2$ and $u_3$) | $D_x(u^{B-1})$ | $D_x(u^{B-1})$ |
| $B-2$ | $u^{B-2}$ (i.e. bit $(B-2)$ of $u_0$, $u_1$, $u_2$ and $u_3$) | $D_x(u^{B-2})$ | $D_x(u^{B-2}) + [D_x(u^{B-1}) \gg 1]$ |
| ... | | | ... |
| ... | | | ... |
| 1 | $u^1$ (i.e. bit 1 of $u_0$, $u_1$, $u_2$ and $u_3$) | $D_x(u^1)$ | $D_x(u^1) + $ (previous contents $\gg 1$) |
| 0 | $u^0$ (i.e. sign bit of $u_0$, $u_1$, $u_2$ and $u_3$) | $D_x(u^0)$ | $-D_x(u^0) + $ (previous contents $\gg 1$) |

clock cycle, one bit from each input $u$ (or each input $v$ if $F_x$ is an odd-numbered coefficient) selects a pre-calculated value $D_x$ from the ROM. The ROM output $D_x$ is added to the previous accumulator contents (right-shifted by 1 bit, equivalent to multiplication by $2^{-1}$). The final output $D_x(u^0)$ is subtracted (this is the sign bit position). After $B$ clock cycles, the accumulator contains the final result $F_x$.

The ROM-Accumulator is a reasonably compact circuit and it is possible to use eight of these in parallel to calculate all eight coefficients $F_x$ of a 1-D FDCT or IDCT in $B$ cycles. The distributed arithmetic design offers good performance with a modest gate count.

There have been examples of multiplier-based DCT designs,[14,15] filter-based designs[12,16] and distributed arithmetic designs.[17,18] A hardware architecture has been presented[19] based on a 'direct' 2-D DCT implementation.

## 7.6   QUANTISATION

In a transform-based video CODEC, the transform stage is usually followed by a quantisa-tion stage. The transforms described in this chapter (DCT and wavelet) are reversible i.e. applying the transform followed by its inverse to image data results in the original image data. This means that the transform process does not remove any information; it simply represents the information in a different form. The quantisation stage removes less 'important' information (i.e. information that does not have a significant influence on the appearance of the reconstructed image), making it possible to compress the remaining data.

In the main image and video coding standards described in Chapters 4 and 5, the quantisation process is split into two parts, an operation in the encoder that converts transform coefficients into *levels* (usually simply described as quantisation) and an operation in the decoder that converts levels into reconstructed transform coefficients (usually described as rescaling or 'inverse quantisation'). The key to this process is that, whilst the original transform coefficients may take on a large number of possible values (like an analogue, 'continuous' signal), the levels and hence the reconstructed coefficients are restricted to a discrete set of values. Figure 7.18 illustrates the quantisation process. Transform coefficients on a continuous scale are quantised to a limited number of possible levels. The levels are rescaled to produce reconstructed coefficients with approximately the same magnitude as the original coefficients *but* a limited number of possible values.

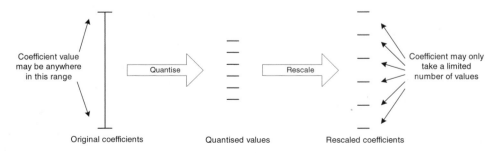

**Figure 7.18**   Quantisation and rescaling

Quantisation has two benefits for the compression process:

1. If the quantisation process is correctly designed, visually significant coefficients (i.e. those that have a significant effect on the quality of the decoded image) are retained (albeit with lower precision), whilst insignificant coefficients are discarded. This typically results in a 'sparse' set of quantised levels, e.g. most of the 64 coefficients in an $8 \times 8$ DCT are set to zero by the quantiser.

2. A sparse matrix containing levels with a limited number of discrete values (the result of quantisation) can be efficiently compressed.

There is, of course, a detrimental effect to image quality because the reconstructed coefficients are not identical to the original set of coefficients and hence the decoded image will not be identical to the original. The amount of compression and the loss of image quality depend on the number of levels produced by the quantiser. A large number of levels means that the coefficient precision is only slightly reduced and compression is low; a small number of levels means a significant reduction in coefficient precision (and image quality) but correspondingly high compression.

*Example*

The DCT coefficients from Table 7.1 are quantised and rescaled with (a) a 'fine' quantiser (with the levels spaced at multiples of 4) and (b) a 'coarse' quantiser (with the levels spaced at multiples of 16). The results are shown in Table 7.4. The finely quantised coefficients (a) retain most of the precision of the originals and 21 non-zero coefficients remain after quantisation. The coarsely quantised coefficients (b) have lost much of their precision and only seven coefficients are left after quantisation (the six coefficients illustrated in Figure 7.6 plus [7, 0]). The finely quantised block will produce a better approximation of the original image block after applying the IDCT; however, the coarsely quantised block will compress to a smaller number of bits.

**Table 7.4** Quantised and rescaled coefficients: (a) fine quantisation, (b) coarse quantisation

| 964 | −4 | −8 | 0 | 0 | 0 | 0 | 0 |
|---|---|---|---|---|---|---|---|
| −160 | −68 | 0 | 4 | 0 | 0 | 0 | 0 |
| 52 | 12 | 0 | 0 | 0 | 0 | 0 | 0 |
| 80 | −8 | −4 | 0 | 0 | 0 | 0 | 0 |
| 36 | −12 | −4 | 0 | 0 | 0 | 0 | 0 |
| −4 | −8 | 0 | 0 | 0 | 0 | 0 | 0 |
| −12 | 0 | 4 | 0 | 0 | 0 | 0 | 0 |
| −20 | 4 | 4 | 0 | 0 | 0 | 0 | 0 |

(a)

| 960 | 0 | 0 | 0 | 0 | 0 | 0 | 0 |
|---|---|---|---|---|---|---|---|
| −160 | −64 | 0 | 0 | 0 | 0 | 0 | 0 |
| 48 | 0 | 0 | 0 | 0 | 0 | 0 | 0 |
| 80 | 0 | 0 | 0 | 0 | 0 | 0 | 0 |
| 32 | 0 | 0 | 0 | 0 | 0 | 0 | 0 |
| 0 | 0 | 0 | 0 | 0 | 0 | 0 | 0 |
| 0 | 0 | 0 | 0 | 0 | 0 | 0 | 0 |
| −16 | 0 | 0 | 0 | 0 | 0 | 0 | 0 |

(b)

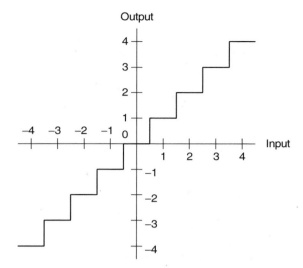

**Figure 7.19**   Linear quantiser

## 7.6.1   Types of Quantiser

The complete quantisation process (forward quantisation followed by rescaling) can be thought of as a mapping between a set of input values and a (smaller) set of output values. The type of mapping has a significant effect on compression and visual quality. Quantisers can be categorised as *linear* or *nonlinear*.

### Linear

The set of input values map to a set of evenly distributed output values and an example is illustrated in Figure 7.19. Plotting the mapping in this way produces a characteristic 'staircase'. A linear quantiser is appropriate when it is required to retain the maximum precision across the entire range of possible input values.

### Nonlinear

The set of output values are not linearly distributed; this means that input values are treated differently depending on their magnitude. A commonly used example is a quantiser with a 'dead zone' about zero, as shown in Figure 7.20. A disproportionately wide range of low-valued inputs are mapped to a zero output. This has the effect of 'favouring' larger values at the expense of smaller ones, i.e. small input values tend to be quantised to zero, whilst larger values are retained. This type of nonlinear quantiser may be used, for example, to quantise 'residual' image data in an inter-coded frame. The residual DCT coefficients (after motion compensation and forward DCT) are distributed about zero. A typical coefficient matrix will contain a large number of near-zero values (positive and negative) and a small number of higher values and a nonlinear quantiser will remove the near-zero values whilst retaining the high values.

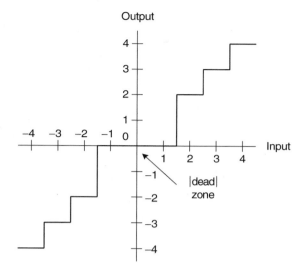

**Figure 7.20**  Nonlinear quantiser with dead zone

Figure 7.21 shows the effect of applying two different nonlinear quantisers to a sine input. The figure shows the input together with the quantised and rescaled output; note the 'dead zone' about zero. The left-hand graph shows a quantiser with 11 levels and the right-hand graph shows a 'coarser' quantiser with only 5 levels.

## 7.6.2  Quantiser Design

The design of the quantisation process has an important effect on compression performance and image quality. The fundamental concepts of quantiser design were presented elsewhere.[20] In order to support compatibility between encoders and decoders, the image and video coding standards specify the *levels* produced by the encoder and the set of *reconstructed coefficients*. However, they do not specify the *forward* quantisation process, i.e. the mapping between the input coefficients and the set of levels. This gives the encoder designer flexibility to design the forward quantiser to give optimum performance for different applications.

For example, the MPEG-4 rescaling process is as follows for inter-coded blocks:

$$|REC| = QUANT \cdot (2 \cdot |LEVEL| + 1) \qquad \text{(if } QUANT \text{ is odd and } LEVEL \neq 0)$$
$$|REC| = QUANT \cdot (2 \cdot |LEVEL| + 1) - 1 \quad \text{(if } QUANT \text{ is even and } LEVEL \neq 0) \quad (7.19)$$
$$REC = 0 \qquad \text{(if } LEVEL = 0)$$

*LEVEL* is the decoded level prior to rescaling and *REC* is the rescaled coefficient. The sign of *REC* is the same as the sign of *LEVEL*. *QUANT* is a quantisation 'scale factor' in the range 1–31. Table 7.5 gives some examples of reconstructed coefficients for a few of the possible combinations of *LEVEL* and *QUANT*. The QUANT parameter controls the *step*

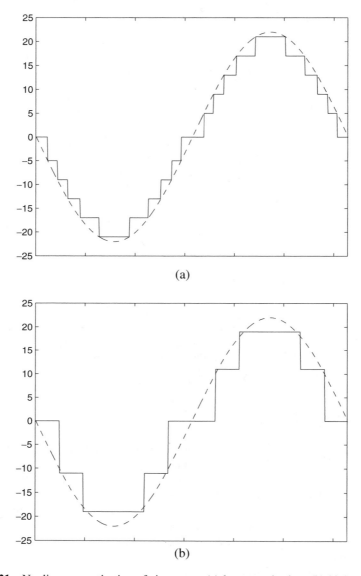

**Figure 7.21**  Nonlinear quantisation of sine wave: (a) low quantisation; (b) high quantisation

*size* of the reconstruction process: outside the 'dead zone' (about zero), the reconstructed values are spaced at intervals of ($QUANT * 2$). A larger value of $QUANT$ means more widely spaced reconstruction levels and this in turn gives higher compression (and poorer decoded image quality).

The other 'half' of the process, the forward quantiser, is not defined by the standard. The design of the forward quantiser determines the range of coefficients ($COEF$) that map to each of the levels. There are many possibilities here: for example, one option is to design the

**Table 7.5** MPEG-4 reconstructed coefficients

| QUANT | ... | −2 | −1 | 0 | 1 | 2 | 3 | 4 | ... |
|-------|-----|-----|-----|---|----|----|----|----|-----|
| | | | | *Levels* | | | | | |
| 1 | ... | −5 | −3 | 0 | 3 | 5 | 7 | 9 | ... |
| 2 | | −9 | −5 | 0 | 5 | 9 | 13 | 17 | |
| 3 | | −15 | −9 | 0 | 9 | 15 | 21 | 27 | |
| 4 | | −19 | −11 | 0 | 11 | 19 | 27 | 35 | |
| ... | | | | | | | | | |

quantiser so that each of the reconstructed values lies at the centre of a range of input values. Figure 7.22 shows an example for $QUANT = 4$. After quantisation and rescaling, original coefficients in the range $(-7 < COEF < 7)$ map to 0 (the 'dead zone'); coefficients in the range $(7 < COEF < 15)$ map to 11; and so on.

However, the quantiser shown in Figure 7.22 is not necessarily the best choice for inter-coded transform coefficients. Figure 7.23 shows the distribution of DCT coefficients in an MPEG-4 coded sequence: most of the coefficients are clustered about zero. Given a quantised coefficient $c'$, the original coefficient $c$ is more likely to have a low value than

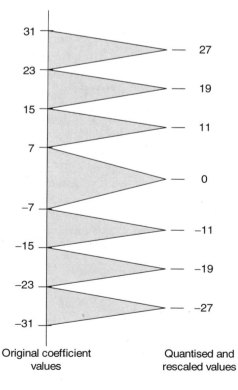

Original coefficient
values

Quantised and
rescaled values

**Figure 7.22** Quantiser (1): *REC* in centre of range

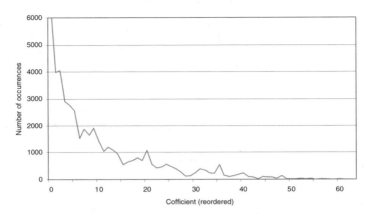

**Figure 7.23** Typical distribution of INTER coefficients

a high value. A better quantiser might 'bias' the reconstructed coefficients towards zero; this means that, on average, the reconstructed values will be closer to the original values (for original coefficient values that are concentrated about zero). An example of a 'biased' forward quantiser design is given in Appendix III of the H.263++ standard:

$$|LEVEL| = (|COEF| - QUANT/2)/(2 * QUANT) \qquad (7.20)$$

The positions of the original and reconstructed coefficients for $QUANT = 4$ are shown in Figure 7.24. Each reconstructed coefficient value ($REC$) is near the lower end of the range of corresponding input coefficients. Inputs in the range ($-10 < COEF < 10$) map to 0; the range ($10 < COEF < 18$) maps to 11, ($-18 < COEF < -10$) maps to $-11$, and so on. For coefficients with a Laplacian probability distribution (i.e. a similar distribution to Figure 7.23), this quantiser design means that, on average, the reconstructed coefficients are closer to the original coefficients. This in turn reduces the error introduced by the quantisation process. Depending on the source video material, the transform coefficient may have different distributions. It is possible to 'redesign' the quantiser (i.e. choose the range that maps to each reconstruction value) to suit a particular set of input data and achieve optimum image quality. This is a computationally intensive process and is a good example of the trade-off between computation and compression performance: applying more computation to the choice of quantiser can lead to better compression performance.

## 7.6.3 Quantiser Implementation

Forward quantisation maps an input coefficient to one of a set of possible levels, depending on the value of a parameter $QUANT$. As Equation 7.20 implies, this can usually be implemented as a division (or as a multiplication by an inverse parameter). The rescaling process (e.g. Equation 7.19) can be implemented as a multiplication.

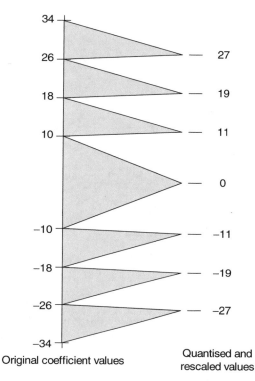

34
26 —— 27
18 —— 19
10 —— 11
      —— 0
-10 —— -11
-18 —— -19
-26 —— -27
-34

Original coefficient values | Quantised and rescaled values

**Figure 7.24** Quantiser (2): *REC* biased towards lower end of range

These operations are reasonably straightforward in software or hardware; however, in some architectures division may be 'expensive', for example requiring many processor cycles in a software CODEC or requiring a large multiplier circuit in a hardware architecture. Rescaling may lend itself to implementation using a look-up table. For example, the MPEG-4 rescaling operation for 'inter' coefficients (Equation 7.19) has a limited number of possible outcomes: there are 31 *QUANT* values and the value of |*LEVEL*| may be in the range 0–128 and so there are $(31 * 129 =)$ 3999 possible reconstructed values. Instead of directly calculating Equation 7.19, the 3999 outcomes may be pre-calculated and stored in a look-up table which the decoder indexes according to *QUANT* and *LEVEL*. This is less practical for the forward quantiser: the magnitude of *COEF* may be in the range 0–2048 and so $(2049 * 31 =)$ 63 519 possible *LEVEL* outcomes need to be stored in the look-up table.

## 7.6.4 Vector Quantisation

In the examples discussed above, each sample (e.g. a transform coefficient) was quantised independently of all other samples (*scalar quantisation*). In contrast, quantising a group of samples as a 'unit' (*vector quantisation*) can offer more scope for efficient compression.[21] In its basic form, vector quantisation is applied in the spatial domain (i.e. it does not involve a transform operation). The heart of a vector quantisation (VQ) CODEC is a *codebook*. This contains a predetermined set of *vectors*, where each vector is a block of samples or pixel. A VQ CODEC operates as follows:

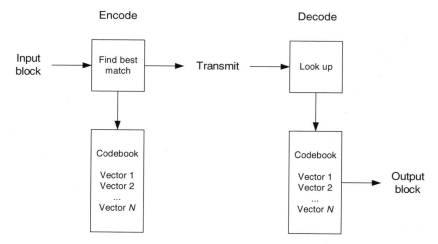

**Figure 7.25**   Vector quantisation CODEC

1. Partition the source image into blocks (e.g. $4 \times 4$ or $8 \times 8$ samples).

2. For each block, choose a vector from the codebook that matches the block as closely as possible.

3. Transmit an *index* that identifies the chosen vector.

4. The decoder extracts the appropriate vector and uses this to represent the original image block.

Figure 7.25 illustrates the operation of a basic VQ CODEC. Compression is achieved by ensuring that the index takes fewer bits to transmit than the original image block itself. VQ is inherently lossy because, for most image blocks, the chosen vector will not be an exact match and hence the reconstructed image block will not be identical to the original. The larger the number of vectors (predetermined image blocks) in the codebook, the higher the likelihood of obtaining a good match. However, a large codebook introduces two difficulties: first, the problem of storing the codebook and second, the problem of searching the codebook to find the optimum match.

The encoder searches the codebook and attempts to minimise the distortion between the original image block $\mathbf{x}$ and the chosen vector $\hat{\mathbf{x}}$, according to some distortion metric (for example, mean squared error: $\|\mathbf{x} - \hat{\mathbf{x}}\|^2$). The search complexity increases with the number of vectors in the codebook $N$, and much of the research into VQ techniques has concentrated on methods of minimising the complexity of the search whilst achieving good compression.

Many modifications to the basic VQ technique described above have been proposed, including the following.

## Tree search VQ

In order to simplify the search procedure in a VQ encoder, the codebook is partitioned into a hierarchy. At each level of the hierarchy, the input image block is compared with just two

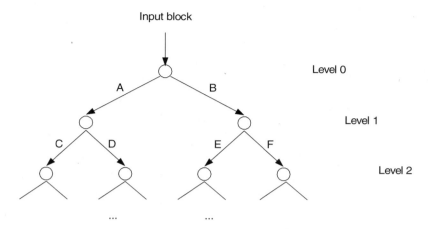

**Figure 7.26** Tree-structured codebook

possible vectors and the best match is chosen. At the next level down, two further choices are offered (based on the choice at the previous level), and so on. Figure 7.26 shows the basic technique: the input block is first compared with two 'root' vectors **A** and **B** (level 0). If **A** is chosen, the next comparison is with vectors **C** and **D**; if **B** is chosen, the next level chooses between **E** and **F**; and so on. In total, $2\log_2 N$ comparisons are required for a codebook of $N$ vectors. This reduction in complexity is offset against a potential loss of image quality compared with a 'full' search of the codebook, since the algorithm is not guaranteed to find the best match out of all possible vectors.

*Variable block size*

In its basic form, with blocks of a constant size, a VQ encoder must transmit an index for every block of the image. Most images have areas of high and low spatial detail and it can be advantageous to use a variable block size for partitioning the image, as shown in Figure 7.27. Prior to quantisation, the image is partitioned into non-overlapping blocks of varying sizes. Small blocks are chosen for areas of the image containing important detail; large blocks are used where there is less detail in the image. Each block is matched with a vector from the codebook and the advantage of this method is that a higher density of vectors (and hence better image reproduction) is achieved for detailed areas of the image, whilst a lower density (and hence fewer transmitted bits) is chosen for less detailed areas. Disadvantages include the extra complexity of the initial partitioning stage and the requirement to transmit a 'map' of the partitioning structure.

*Practical considerations*

Vector quantisation is highly asymmetrical in terms of computational complexity. Encoding involves an intensive search operation for every image block, whilst decoding involves a simple table look-up. VQ (in its basic form) is therefore unsuitable for many two-way video

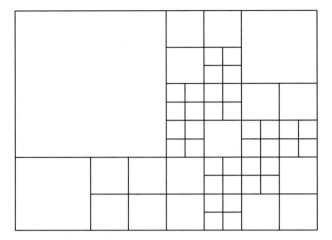

**Figure 7.27**    Image partitioned into varying block sizes

communication applications but attractive for applications where low decoder complexity is required. At present, VQ has not found its way into any of the 'mainstream' video and image coding standards. However, it continues to be an active area for research and increasingly sophisticated techniques (such as fast codebook search algorithms and VQ combined with other image coding methods) may lead to increased uptake in the future.

## 7.7  SUMMARY

The most popular method of compressing images (or motion-compensated residual frames) is by applying a transform followed by quantisation. The purpose of an image transform is to decorrelate the original image data and to 'compact' the energy of the image. After decorrelation and compaction, most of the image energy is concentrated into a small number of coefficients which are 'clustered' together.

The DCT is usually applied to $8 \times 8$ blocks of image or residual data. The basic 2-D transform is relatively complex to implement, but the computation can be significantly reduced first by splitting it into two 1-D transforms and second by exploiting symmetry properties to simplify each 1-D transform. 'Flowgraph'-type fast algorithms are suitable for software implementations and a range of algorithms enable the designer to tailor the choice of FDCT to the processing platform. The more regular parallel-multiplier or distributed arithmetic algorithms are better suited to dedicated hardware designs.

The design of the quantiser can have an important contribution to image quality in an image or video CODEC. After quantisation, the remaining significant transform coefficients are entropy encoded together with side information (such as headers and motion vectors) to form a compressed representation of the original image or video sequence. The next chapter will examine the theory and practice of designing efficient entropy encoders and decoders.

# REFERENCES

1. N. Ahmed, T. Natrajan and K. R. Rao, 'Discrete cosine transform', *IEEE Trans. Computers*, January 1974.
2. W. A. Pearlman, 'Trends of tree-based, set-partitioning compression techniques in still and moving image systems', *Proc. PCS01*, Seoul, April 2001.
3. W-H. Chen, C. H. Smith and S. C. Fralick, 'A fast computational algorithm for the discrete cosine transform', *IEEE Trans. Communications*, **COM-25**, No. 9, September 1977.
4. B. G. Lee, 'A new algorithm to compute the discrete cosine transform', *IEEE Trans. ASSP*, **32**(6), December 1984.
5. C. Loeffler, A. Ligtenberg and G. Moschytz, 'Practical fast 1-D DCT algorithms with 11 multiplications', *Proc. ICASSP-89*, 1989.
6. Y. Arai, T. Agui and M. Nakajima, 'A fast DCT-SQ scheme for images', *Trans. of the IEICE E*, **71**(11), November 1988.
7. M. Vetterli and H. Nussbaumer, 'Simple FFT and DCT algorithms with reduced number of operations', *Signal Processing*, **6**(4), August 1984.
8. E. Feig and S. Winograd, 'Fast algorithms for the discrete cosine transform', *IEEE Trans. Signal Processing*, **40**(9), September 1992.
9. F. A. Kamangar and K. A. Rao, 'Fast algorithms for the 2-D discrete cosine transform', *IEEE Trans. Computers*, **31**(9), September 1982.
10. M. Vetterli, 'Fast 2-D discrete cosine transform', *Proc. IEEE ICASSP*, 1985.
11. A. Peled and B. Liu, 'A new hardware realization of digital filters', *IEEE Trans. ASSP*, **22**(6), December 1974.
12. J. R. Spanier, G. Keane, J. Hunter and R. Woods, 'Low power implementation of a discrete cosine transform IP Core', *Proc. DATE-2000*, Paris, March 2000.
13. T. D. Tran, 'The BinDCT: fast multiplierless approximation of the DCT', *IEEE Signal Processing Letters*, **7**, June 2000.
14. M. Sanchez, J. Lopez, O. Plata, M. Trenas and E. Zapata, 'An efficient architecture for the in-place fast cosine transform', *J. VLSI Sig. Proc.*, **21**(2), June 1999.
15. G. Aggarwal and D. Gajski, *Exploring DCT Implementations*, UC Irvine Tech Report TR-98-10, March 1998.
16. G. A. Jullien, 'VLSI digital signal processing: some arithmetic issues', *Proc. SPIE*, **2846**, Advanced Signal Processing Algorithms, Architectures and Implementations, October 1996.
17. M. T. Sun, T. C. Chen and A. Gottlieb, 'VLSI implementation of a $16 * 16$ discrete cosine transform', *IEEE Trans. Circuits and Systems*, **36**(4), April 1989.
18. T-S. Chang, C-S. Kung and C-W. Jen, 'A simple processor core design for DCT/IDCT', *IEEE Trans. on CSVT*, **10**(3), April 2000.
19. P. Lee and G. Liu, 'An efficient algorithm for the 2D discrete cosine transform', *Signal Processing*, **55**(2), Dec. 1996, pp. 221–239.
20. Y. Shoham and A. Gersho, 'Efficient bit allocation for an arbitrary set of quantisers', *IEEE Trans. ACSSP*, **32**(3), June 1984.
21. N. Nasrabadi and R. King, 'Image coding using vector quantisation: a review', *IEEE Trans. Communications*, **36**(8), August 1988.

# 8

# Entropy Coding

## 8.1 INTRODUCTION

A video encoder contains two main functions: a *source model* that attempts to represent a video scene in a compact form that is easy to compress (usually an approximation of the original video information) and an *entropy encoder* that compresses the output of the model prior to storage and transmission. The source model is matched to the characteristics of the input data (images or video frames), whereas the entropy coder may use 'general-purpose' statistical compression techniques that are not necessarily unique in their application to image and video coding.

As with the functions described earlier (motion estimation and compensation, transform coding, quantisation), the design of an entropy CODEC is affected by a number of constraints including:

1. *Compression efficiency*: the aim is to represent the source model output using as few bits as possible.

2. *Computational efficiency*: the design should be suitable for implementation on the chosen hardware or software platform.

3. *Error robustness*: if transmission errors are likely, the entropy CODEC should support recovery from errors and should (if possible) limit error propagation at decoder (this constraint may conflict with (1) above).

In a typical transform-based video CODEC, the data to be encoded by the entropy CODEC falls into three main categories: transform coefficients (e.g. quantised DCT coefficients), motion vectors and 'side' information (headers, synchronisation markers, etc.). The method of coding side information depends on the standard. Motion vectors can often be represented compactly in a differential form due to the high correlation between vectors for neighbouring blocks or macroblocks. Transform coefficients can be represented efficiently with 'run–level' coding, exploiting the sparse nature of the DCT coefficient array.

An entropy encoder maps input symbols (for example, run–level coded coefficients) to a compressed data stream. It achieves compression by exploiting redundancy in the set of input symbols, representing frequently occurring symbols with a small number of bits and infrequently occurring symbols with a larger number of bits. The two most popular entropy encoding methods used in video coding standards are Huffman coding and arithmetic coding. Huffman coding (or 'modified' Huffman coding) represents each input symbol by a variable-length codeword containing an integral number of bits. It is relatively straightforward to implement, but cannot achieve optimal compression because of the restriction that each codeword must contain an integral number of bits. Arithmetic coding maps an input symbol into a fractional number of bits, enabling greater compression efficiency at the expense of higher complexity (depending on the implementation).

## 8.2 DATA SYMBOLS

### 8.2.1 Run–Level Coding

The output of the quantiser stage in a DCT-based video encoder is a block of quantised transform coefficients. The array of coefficients is likely to be sparse: if the image block has been efficiently decorrelated by the DCT, most of the quantised coefficients in a typical block are zero. Figure 8.1 shows a typical block of quantised coefficients from an MPEG-4 'intra' block. The structure of the quantised block is fairly typical. A few non-zero coefficients remain after quantisation, mainly clustered around DCT coefficient $(0, 0)$: this is the 'DC' coefficient and is usually the most important coefficient to the appearance of the reconstructed image block.

The block of coefficients shown in Figure 8.1 may be efficiently compressed as follows:

1. *Reordering*. The non-zero values are clustered around the top left of the 2-D array and this stage groups these non-zero values together.

2. *Run–level coding*. This stage attempts to find a more efficient representation for the large number of zeros (48 in this case).

3. *Entropy coding*. The entropy encoder attempts to reduce the redundancy of the data symbols.

### Reordering

The optimum method of reordering the quantised data depends on the distribution of the non-zero coefficients. If the original image (or motion-compensated residual) data is evenly

DC coefficient

| 102 | -33 | -3 | -4 | -2 | -1 | | |
|---|---|---|---|---|---|---|---|
| 21 | -2 | -3 | | -1 | | | |
| -3 | | 1 | | | | | |
| 2 | | | | | | | |
| 1 | | 1 | | | | | |
| -2 | | | | | | | |
| | | | | | | | |
| | | | | | | | |

**Figure 8.1** Block of quantised coefficients (intra-coding)

distributed in the horizontal and vertical directions (i.e. there is not a predominance of 'strong' image features in either direction), then the significant coefficients will also tend to be evenly distributed about the top left of the array (Figure 8.2(a)). In this case, a zigzag reordering pattern such as Figure 8.2 (c) should group together the non-zero coefficients

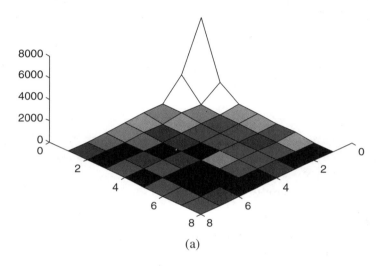

(a)

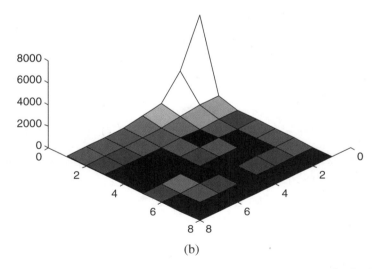

(b)

**Figure 8.2** Typical data distributions and reordering patterns: (a) even distribution; (b) field distribution; (c) zigzag; (d) modified zigzag

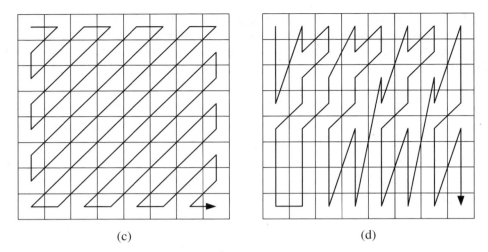

(c)                                                         (d)

**Figure 8.2**   (*Continued*)

efficiently. However, in some cases an alternative pattern performs better. For example, a *field* of interlaced video tends to vary more rapidly in the vertical than in the horizontal direction (because it has been vertically subsampled). In this case the non-zero coefficients are likely to be 'skewed' as shown in Figure 8.2(b): they are clustered more to the left of the array (corresponding to basis functions with a strong vertical variation, see for example Figure 7.4). A modified reordering pattern such as Figure 8.2(d) should perform better at grouping the coefficients together.

*Run–level coding*

The output of the reordering process is a linear array of quantised coefficients. Non-zero coefficients are mainly grouped together near the start of the array and the remaining values in the array are zero. Long sequences of identical values (zeros in this case) can be represented as a (run, level) code, where (run) indicates the number of zeros preceding a non-zero value and (level) indicates the sign and magnitude of the non-zero coefficient.

The following example illustrates the reordering and run–level coding process.

*Example*

The block of coefficients in Figure 8.1 is reordered with the zigzag scan shown in Figure 8.2 and the reordered array is run–level coded.

Reordered array:
[102, −33, 21, −3, −2, −3, −4, −3, 0, 2, 1, 0, 1, 0, −2, −1, −1, 0, 0, 0, −2, 0, 0, 0, 0, 0, 0, 0, 0, 0, 0, 0, 1, 0 …]

Run–level coded:
(0, 102) (0, −33) (0, 21) (0, −3) (0, −2) (0, −3) (0, −4) (0, −3) (1, 2) (0, 1) (1, 1) (1, −2) (0, −1) (0, −1) (4, −2) (11, 1)

Two special cases need to be considered. Coefficient (0, 0) (the 'DC' coefficient) is important to the appearance of the reconstructed image block and has no preceding zeros. In an intra-coded block (i.e. coded without motion compensation), the DC coefficient is rarely zero and so is treated differently from other coefficients. In an H.263 CODEC, intra-DC coefficients are encoded with a fixed, relatively low quantiser setting (to preserve image quality) and without (run, level) coding. Baseline JPEG takes advantage of the property that neighbouring image blocks tend to have similar mean values (and hence similar DC coefficient values) and each DC coefficient is encoded differentially from the previous DC coefficient.

The second special case is the final run of zeros in a block. Coefficient (7, 7) is usually zero and so we need a special case to deal with the final run of zeros that has no terminating non-zero value. In H.261 and baseline JPEG, a special code symbol, 'end of block' or EOB, is inserted after the last (run, level) pair. This approach is known as 'two-dimensional' run–level coding since each code represents just two values (run and level). The method does not perform well under high compression: in this case, many blocks contain only a DC coefficient and so the EOB codes make up a significant proportion of the coded bit stream. H.263 and MPEG-4 avoid this problem by encoding a flag along with each (run, level) pair. This 'last' flag signifies the final (run, level) pair in the block and indicates to the decoder that the rest of the block should be filled with zeros. Each code now represents three values (run, level, last) and so this method is known as 'three-dimensional' run–level-last coding.

## 8.2.2 Other Symbols

In addition to run–level coded coefficient data, a number of other values need to be coded and transmitted by the video encoder. These include the following.

### Motion vectors

The vector displacement between the current and reference areas (e.g. macroblocks) is encoded along with each data unit. Motion vectors for neighbouring data units are often very similar, and this property may be used to reduce the amount of information required to be encoded. In an H.261 CODEC, for example, the motion vector for each macroblock is predicted from the preceding macroblock. The difference between the current and previous vector is encoded and transmitted (instead of transmitting the vector itself). A more sophisticated prediction is formed during MPEG-4/H.263 coding: the vector for each macroblock (or block if the optional advanced prediction mode is enabled) is predicted from up to three previously transmitted motion vectors. This helps to further reduce the transmitted information. These two methods of predicting the current motion vector are shown in Figure 8.3.

### Example

| | | |
|---|---|---|
| Motion vector of current macroblock: | $x = +3.5,$ | $y = +2.0$ |
| Predicted motion vector from previous macroblocks: | $x = +3.0,$ | $y = 0.0$ |
| Differential motion vector: | $dx = +0.5,$ | $dy = -2.0$ |

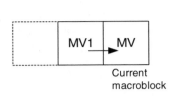

H.261: predict MV from previous
macroblock vector MV1

H.263/MPEG4: predict MV from three previous
macroblock vectors MV1, MV2 and MV3

**Figure 8.3**   Motion vector prediction (H.261, H.263)

*Quantisation parameter*

In order to maintain a target bit rate, it is common for a video encoder to modify the quantisation parameter (scale factor or step size) during encoding. The change must be signalled to the decoder. It is not usually desirable to suddenly change the quantisation parameter by a large amount during encoding of a video frame and so the parameter may be encoded differentially from the previous quantisation parameter.

*Flags to indicate presence of coded units*

It is common for certain components of a macroblock not to be present. For example, efficient motion compensation and/or high compression leads to many blocks containing only zero coefficients after quantisation. Similarly, macroblocks in an area that contains no motion or homogeneous motion will tend to have zero motion vectors (after differential prediction as described above). In some cases, a macroblock may contain no coefficient data and a zero motion vector, i.e. nothing needs to be transmitted. Rather than encoding and sending zero values, it can be more efficient to encode flag(s) that indicate the presence or absence of these data units.

*Example*

Coded block pattern (CBP) indicates the blocks containing non-zero coefficients in an inter-coded macroblock.

| Number of non-zero coefficients in each block | | | | | | |
|---|---|---|---|---|---|---|
| Y0 | Y1 | Y2 | Y3 | Cr | Cb | CBP |
| 2 | 1 | 0 | 7 | 0 | 0 | 110100 |
| 0 | 6 | 9 | 1 | 1 | 3 | 011111 |

*Synchronisation markers*

A video decoder may require to resynchronise in the event of an error or interruption to the stream of coded data. Synchronisation markers in the bit stream provide a means of doing this. Typically, the differential predictions mentioned above (DC coefficient, motion vectors and quantisation parameter) are reset after a synchronisation marker, so that the data after the marker may be decoded independently of previous (perhaps errored) data. Synchronisation is supported by restart markers in JPEG, group of block (GOB) headers in baseline H.263 and MPEG-4 (at fixed intervals within the coded picture) and slice start codes in the MPEG-1, MPEG-2 and annexes to H.263 and MPEG-4 (at user definable intervals).

*Higher-level headers*

Information that applies to a complete frame or picture is encoded in a header (picture header). Higher-level information about a sequence of frames may also be encoded (for example, sequence and group of pictures headers in MPEG-1 and MPEG-2).

## 8.3 HUFFMAN CODING

A Huffman entropy encoder maps each input symbol into a variable length codeword and this type of coder was first proposed in 1952.[1] The constraints on the variable length codeword are that it must (a) contain an integral number of bits and (b) be uniquely decodeable (i.e. the decoder must be able to identify each codeword without ambiguity).

### 8.3.1 'True' Huffman Coding

In order to achieve the maximum compression of a set of data symbols using Huffman encoding, it is necessary to calculate the probability of occurrence of each symbol. A set of variable length codewords is then constructed for this data set. This process will be illustrated by the following example.

*Example: Huffman coding, 'Carphone' motion vectors*

A video sequence, 'Carphone', was encoded with MPEG-4 (short header mode). Table 8.1 lists the probabilities of the most commonly occurring motion vectors in the encoded

**Table 8.1** Probability of occurrence of motion vectors in 'Carphone' sequence

| Vector | Probability $P$ | $\log_2(1/P)$ |
|---|---|---|
| −1.5 | 0.014 | 6.16 |
| −1 | 0.024 | 5.38 |
| −0.5 | 0.117 | 3.10 |
| 0 | 0.646 | 0.63 |
| 0.5 | 0.101 | 3.31 |
| 1 | 0.027 | 5.21 |
| 1.5 | 0.016 | 5.97 |

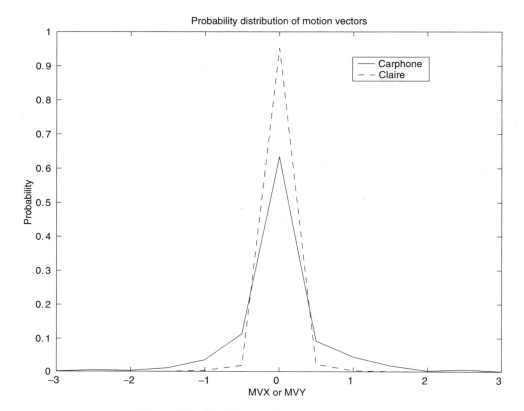

**Figure 8.4**   Distribution of motion vector values

sequence and their *information content*, $\log_2(1/P)$. To achieve optimum compression, each value should be represented with exactly $\log_2(1/P)$ bits.

The vector probabilities are shown graphically in Figure 8.4 (the solid line). '0' is the most common value and the probability drops sharply for larger motion vectors. (Note that there are a small number of vectors larger than $+/-1.5$ and so the probabilities in the table do not sum to 1.)

## 1. Generating the Huffman code tree

To generate a Huffman code table for this set of data, the following iterative procedure is carried out (we will ignore any vector values that do not appear in Table 8.1):

1. Order the list of data in increasing order of probability.

2. Combine the two lowest-probability data items into a 'node' and assign the joint probability of the data items to this node.

3. Reorder the remaining data items and node(s) in increasing order of probability and repeat step 2.

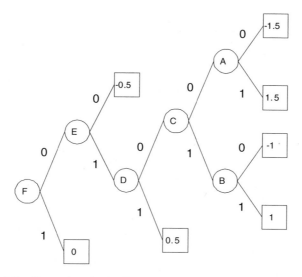

**Figure 8.5**  Generating the Huffman code tree: 'Carphone' motion vectors

The procedure is repeated until there is a single 'root' node that contains all other nodes and data items listed 'beneath' it. This procedure is illustrated in Figure 8.5.

- *Original list*: The data items are shown as square boxes. Vectors ($-1.5$) and (1.5) have the lowest probability and these are the first candidates for merging to form node 'A'.

- *Stage 1*: The newly created node 'A' (shown as a circle) has a probability of 0.03 (from the combined probabilities of ($-1.5$) and (1.5)) and the two lowest-probability items are vectors ($-1$) and (1). These will be merged to form node 'B'.

- *Stage 2*: A and B are the next candidates for merging (to form 'C').

- *Stage 3*: Node C and vector (0.5) are merged to form 'D'.

- *Stage 4*: ($-0.5$) and D are merged to form 'E'.

- *Stage 5*: There are two 'top-level' items remaining: node E and the highest-probability vector (0). These are merged to form 'F'.

- *Final tree*: The data items have all been incorporated into a binary 'tree' containing seven data values and six nodes. Each data item is a 'leaf' of the tree.


## 2. Encoding

Each 'leaf' of the binary tree is mapped to a VLC. To find this code, the tree is 'traversed' from the root node (F in this case) to the leaf (data item). For every branch, a 0 or 1 is appended to the code: 0 for an upper branch, 1 for a lower branch (shown in the final tree of Figure 8.5). This gives the following set of codes (Table 8.2). Encoding is achieved by transmitting the appropriate code for each data item. Note that once the tree has been generated, the codes may be stored in a look-up table.

**Table 8.2**  Huffman codes: 'Carphone' motion vectors

| Vector | Code  | Bits (actual) | Bits (ideal) |
|--------|-------|---------------|--------------|
| 0      | 1     | 1             | 0.63         |
| −0.5   | 00    | 2             | 3.1          |
| 0.5    | 011   | 3             | 3.31         |
| −1.5   | 01000 | 5             | 6.16         |
| 1.5    | 01001 | 5             | 5.97         |
| −1     | 01010 | 5             | 5.38         |
| 1      | 01011 | 5             | 5.21         |

Note the following points:

1. High probability data items are assigned short codes (e.g. 1 bit for the most common vector '0'). However, the vectors (−1.5, 1.5, −1, 1) are each assigned 5-bit codes (despite the fact that −1 and −1 have higher probabilities than 1.5 and 1.5). The lengths of the Huffman codes (each an integral number of bits) do not match the 'ideal' lengths given by $\log_2(1/P)$.

2. No code contains any other code as a prefix, i.e. reading from the left-hand bit, each code is uniquely decodable.

For example, the series of vectors (1, 0, 0.5) would be transmitted as follows:

$$01011|1|011$$

## 3. Decoding

In order to decode the data, the decoder must have a local copy of the Huffman code tree (or look-up table). This may be achieved by transmitting the look-up table itself, or sending the list of data and probabilities, prior to sending the coded data. Each uniquely decodable code may then be read and converted back to the original data. Following the example above:

01011 is decoded as (1)
1 is decoded as (0)
011 is decoded as (0.5)

## Example: Huffman coding, 'Claire' motion vectors

Repeating the process described above for the video sequence 'Claire' gives a different result. This sequence contains less motion than 'Carphone' and so the vectors have a different distribution (shown in Figure 8.4, dotted line). A much higher proportion of vectors are zero (Table 8.3).

The corresponding Huffman tree is given in Figure 8.6. Note that the 'shape' of the tree has changed (because of the distribution of probabilities) and this gives a different set of

**Table 8.3** Probability of occurrence of motion vectors in 'Claire' sequence

| Vector | Probability | $\log_2(1/P)$ |
|--------|-------------|---------------|
| $-1.5$ | 0.001 | 9.66 |
| $-1$ | 0.003 | 8.38 |
| $-0.5$ | 0.018 | 5.80 |
| 0 | 0.953 | 0.07 |
| 0.5 | 0.021 | 5.57 |
| 1 | 0.003 | 8.38 |
| 1.5 | 0.001 | 9.66 |

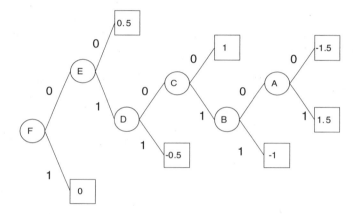

**Figure 8.6** Huffman tree for 'Claire' motion vectors

Huffman codes (shown in Table 8.4). There are still six nodes in the tree, one less than the number of data items (seven): this is always the case with Huffman coding.

If the probability distributions are accurate, Huffman coding provides a relatively compact representation of the original data. In these examples, the frequently occurring (0) vector is represented very efficiently as a single bit. However, to achieve optimum compression, a

**Table 8.4** Huffman codes: 'Claire' motion vectors

| Vector | Code | Bits (actual) | Bits (ideal) |
|--------|------|---------------|--------------|
| 0 | 1 | 1 | 0.07 |
| 0.5 | 00 | 2 | 5.57 |
| $-0.5$ | 011 | 3 | 5.8 |
| 1 | 0100 | 4 | 8.38 |
| $-1$ | 01011 | 5 | 8.38 |
| $-1.5$ | 010100 | 6 | 9.66 |
| 1.5 | 010101 | 6 | 9.66 |

separate code table is required for each of the two sequences 'Carphone' and 'Claire'. The loss of potential compression efficiency due to the requirement for integral length codes is very obvious for vector '0' in the 'Claire' sequence: the optimum number of bits (information content) is 0.07 but the best that can be achieved with Huffman coding is 1 bit.

### 8.3.2   Modified Huffman Coding

The Huffman coding process described above has two disadvantages for a practical video CODEC. First, the decoder must use the same codeword set as the encoder. This means that the encoder needs to transmit the information contained in the probability table before the decoder can decode the bit stream, an extra overhead that reduces compression efficiency. Second, calculating the probability table for a large video sequence (prior to generating the Huffman tree) is a significant computational overhead and cannot be done until after the video data is encoded. For these reasons, the image and video coding standards define sets of codewords based on the probability distributions of a large range of video material. Because the tables are 'generic', compression efficiency is lower than that obtained by pre-analysing the data to be encoded, especially if the sequence statistics differ significantly from the 'generic' probability distributions. The advantage of not requiring to calculate and transmit individual probability tables usually outweighs this disadvantage. (Note: Annex C of the original JPEG standard supports individually calculated Huffman tables, but most practical implementations use the 'typical' Huffman tables provided in Annex K of the standard.)

### 8.3.3   Table Design

The following two examples of VLC table design are taken from the H.263 and MPEG-4 standards. These tables are required for H.263 'baseline' coding and MPEG-4 'short video header' coding.

*H.263/MPEG-4 transform coefficients (TCOEF)*

H.263 and MPEG-4 use '3-dimensional' coding of quantised coefficients, where each codeword represents a combination of (run, level, last) as described in Section 8.2.1. A total of 102 specific combinations of (run, level, last) have VLCs assigned to them. Table 8.5 shows 26 of these codes.

A further 76 VLCs are defined, each up to 13 bits long. Note that the last bit of each codeword is the sign bit 's', indicating the sign of the decoded coefficient (0 = positive, 1 = negative). Any (run, level, last) combination that is not listed in the table is coded using an escape sequence, a special ESCAPE code (0000011) followed by a 13-bit fixed length code describing the values of run, level and last.

The codes shown in Table 8.5 are represented in 'tree' form in Figure 8.7. A codeword containing a run of more than eight zeros is not valid, so any codeword starting with 000000000 . . . indicates an error in the bit stream (or possibly a start code, which begins with a long sequence of zeros, occurring at an unexpected position in the sequence). All other sequences of bits can be decoded as valid codes. Note that the smallest codes are

**Table 8.5**  H.263/MPEG4 transform coefficient (TCOEF)
VLCs (partial, all codes < 9 bits)

| Last | Run | Level | Code |
|---|---|---|---|
| 0 | 0 | 1 | 10s |
| 0 | 1 | 1 | 110s |
| 0 | 2 | 1 | 1110s |
| 0 | 0 | 2 | 1111s |
| 1 | 0 | 1 | 0111s |
| 0 | 3 | 1 | 01101s |
| 0 | 4 | 1 | 01100s |
| 0 | 5 | 1 | 01011s |
| 0 | 0 | 3 | 010101s |
| 0 | 1 | 2 | 010100s |
| 0 | 6 | 1 | 010011s |
| 0 | 7 | 1 | 010010s |
| 0 | 8 | 1 | 010001s |
| 0 | 9 | 1 | 010000s |
| 1 | 1 | 1 | 001111s |
| 1 | 2 | 1 | 001110s |
| 1 | 3 | 1 | 001101s |
| 1 | 4 | 1 | 001100s |
| 0 | 0 | 4 | 0010111s |
| 0 | 10 | 1 | 0010110s |
| 0 | 11 | 1 | 0010101s |
| 0 | 12 | 1 | 0010100s |
| 1 | 5 | 1 | 0010011s |
| 1 | 6 | 1 | 0010010s |
| 1 | 7 | 1 | 0010001s |
| 1 | 8 | 1 | 0010000s |
| ESCAPE | | | 0000011s |
| ... | ... | ... | ... |

allocated to short runs and small levels (e.g. code '10' represents a run of 0 and a level of
$+/-1$), since these occur most frequently.

## H.263/MPEG-4 motion vector difference (MVD)

The H.263/MPEG-4 differentially coded motion vectors (MVD) described in Section 8.2.2
are each encoded as a pair of VLCs, one for the x-component and one for the y-component.
Part of the table of VLCs is shown in Table 8.6 and in 'tree' form in Figure 8.8. A further
49 codes (8–13 bits long) are not shown here. Note that the shortest codes represent small
motion vector differences (e.g. MVD $=0$ is represented by a single bit code '1').

## H.26L universal VLC (UVLC)

The emerging H.26L standard takes a step away from individually calculated Huffman tables
by using a 'universal' set of VLCs for any coded element. Each codeword is generated from

176

ENTROPY CODING

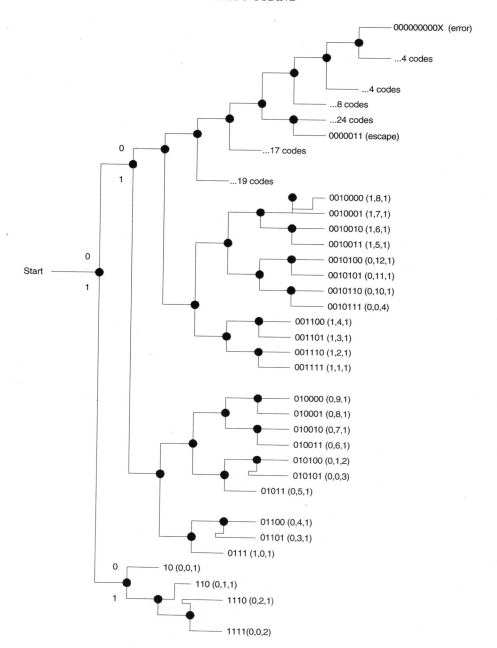

**Figure 8.7** H.263/MPEG-4 TCOEF VLCs

**Table 8.6** H.263/MPEG-4 motion vector difference (MVD) VLCs

| MVD | Code |
|-----|------|
| 0 | 1 |
| +0.5 | 010 |
| −0.5 | 011 |
| +1 | 0010 |
| −1 | 0011 |
| +1.5 | 00010 |
| −1.5 | 00011 |
| +2 | 0000110 |
| −2 | 0000111 |
| +2.5 | 00001010 |
| −2.5 | 00001011 |
| +3 | 00001000 |
| −3 | 00001001 |
| +3.5 | 00000110 |
| −3.5 | 00000111 |
| . . . | . . . |

the following systematic list:

$$1$$
$$0\ x_0\ 1$$
$$0\ x_1\ 0\ x_0\ 1$$
$$0\ x_2\ 0\ x_1\ 0\ x_0\ 1$$
$$0\ x_3\ 0\ x_2\ 0\ x_1\ 0\ x_0\ 1$$
$$. . .$$

where $x_k$ is a single bit. Hence there is one 1-bit codeword; two 3-bit codewords; four 5-bit codewords; eight 7-bit codewords; and so on. Table 8.7 shows the first 12 codes and these are represented in tree form in Figure 8.9. The highly regular structure of the set of codewords can be seen in this figure.

Any data element to be coded (transform coefficients, motion vectors, block patterns, etc.) is assigned a code from the list of UVLCs. The codes are not optimised for a specific data element (since the same set of codes is used for all elements): however, the uniform, regular structure considerably simplifies encoder and decoder design since the same methods can be used to encode or decode any data element.

### 8.3.4 Entropy Coding Example

This example follows the process of encoding and decoding a block of quantised coefficients in an MPEG-4 inter-coded picture. Only six non-zero coefficients remain in the block: this

**Table 8.7**   H.26L universal VLCs

| Index | $x_2$ | $x_1$ | $x_0$ | Codeword |
|-------|-------|-------|-------|----------|
| 0     |       |       | N/A   | 1        |
| 1     |       |       | 0     | 001      |
| 2     |       |       | 1     | 011      |
| 3     |       | 0     | 0     | 00001    |
| 4     |       | 0     | 1     | 00011    |
| 5     |       | 1     | 0     | 01001    |
| 6     |       | 1     | 1     | 01011    |
| 7     | 0     | 0     | 0     | 0000001  |
| 8     | 0     | 0     | 1     | 0000011  |
| 9     | 0     | 1     | 0     | 0001001  |
| 10    | 0     | 1     | 1     | 0001011  |
| 11    | 1     | 0     | 0     | 0100001  |
| . . . | . . . | . . . | . . . | . . .    |

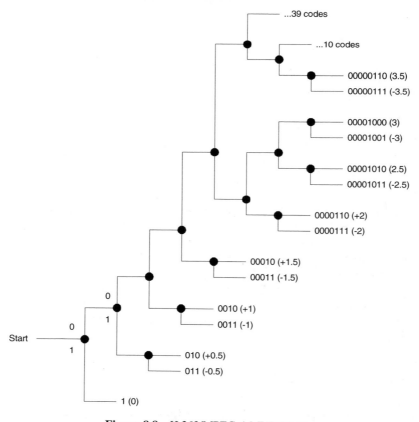

**Figure 8.8**   H.263/MPEG-4 MVD VLCs

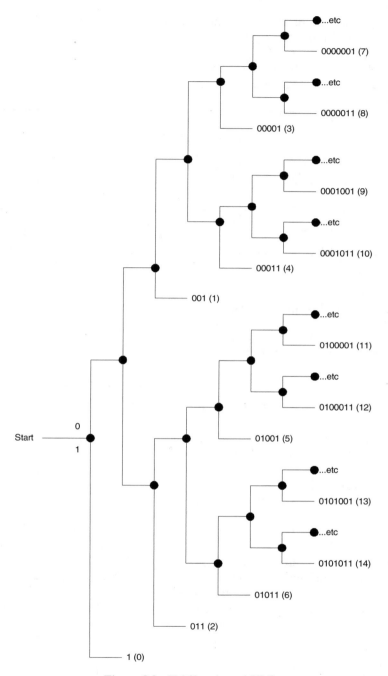

**Figure 8.9** H.26L universal VLCs

would be characteristic of either a highly compressed block or a block that has been efficiently predicted by motion estimation.

Quantised DCT coefficients (empty cells are '0'):

| 4  | -1 |   |   | 1 |   |   |   |
|----|----|---|---|---|---|---|---|
|    | -3 |   |   |   |   |   |   |
| 2  |    |   |   |   |   |   |   |
|    |    |   |   |   |   |   |   |
| -1 |    |   |   |   |   |   |   |
|    |    |   |   |   |   |   |   |
|    |    |   |   |   |   |   |   |
|    |    |   |   |   |   |   |   |

Zigzag reordered coefficients:

| 4 | -1 | 0 | 2 | -3 | 0 | 0 | 0 | 0 | 0 | -1 | 0 | 0 | 0 | 1 | 0 | 0 | ... |
|---|----|---|---|----|---|---|---|---|---|----|---|---|---|---|---|---|-----|

(last, run, level):
(0, 0, 4); (0, 0, -1); (0, 1, 2); (0, 0, -3); (0, 5, -1); (1, 3, 1)

TCOEF variable length codes: (from Table 8.5: note that the last bit is the sign)
00101110; 101; 0101000; 0101011; 010111; 0011010

Transmitted bit sequence:
0010111010101010000101011010111 0011010

Decoding of this sequence proceeds as follows. The decoder 'steps' through the TCOEF tree (shown in Figure 8.7) until it reaches the 'leaf' 0010111. The next bit (0) is decoded as the sign and the (last, run, level) group (0, 0, 4) is obtained. The steps taken by the decoder for this first coefficient are highlighted in Figure 8.10. The process is repeated with the 'leaf' 10 followed by sign (1) and so on until a 'last' coefficient is decoded. The decoder can now fill the coefficient array and reverse the zigzag scan to restore the array of $8 \times 8$ quantised coefficients.

## 8.3.5   Variable Length Encoder Design

*Software design*

A general approach to variable-length encoding in software is as follows:

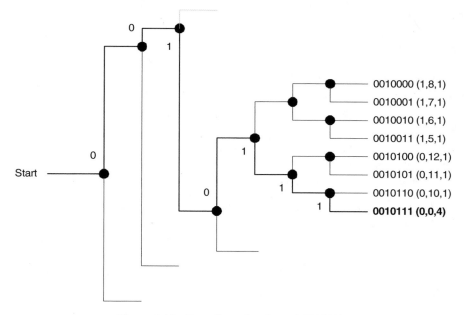

**Figure 8.10**   Decoding of codeword 0010111s

```
for each data symbol
    find the corresponding VLC value and length (in bits)
    pack this VLC into an output register R
    if the contents of R exceed L bytes
        write L (least significant) bytes to the output stream
        shift R by L bytes
```

*Example*

Using the entropy encoding example above, $L = 1$ byte, $R$ is empty at start of encoding:

| VLC | \multicolumn{2}{c}{$R$ (before output)} | | \multicolumn{2}{c}{$R$ (after output)} | |
|---|---|---|---|---|
| | Size | Value | Size | Value |
| 00101110 | 8 | **00101110** | 0 | — |
| 101 | 3 | **101** | | |
| 0101000 | 10 | **0101000101** | 2 | 01 |
| 0101011 | 9 | **010101101** | 1 | 0 |
| 010111 | 7 | **0101111** | | |
| 0011010 | 14 | **00110100101111** | 6 | 001101 |

The following packed bytes are written to the output stream: 00101110, 01000101, 10101101, 00101111. At the end of the above sequence, the output register $R$ still contains 6 bits (001101). If encoding stops here, it will be necessary to 'flush' the contents of $R$ to the output stream.

The MVD codes listed in Table 8.6 can be stored in a simple look-up table. Only 64 valid MVD values exist and the contents of the look-up table are as follows:

$$[\text{ index }][\text{vlc}][\text{ length }]$$

where [index] is a number in the range $0 \ldots 63$ that is derived directly from MVD, [vlc] is the variable length code 'padded' with zeros and represented with a fixed number of bits (e.g. 16 or 32 bits) and [length] indicates the number of bits present in the variable length code.

Converting (last, run, level) into the TCOEF VLCs listed in Table 8.5 is slightly more problematic. The 102 predetermined combinations of (last, run, level) have individual VLCs assigned to them (these are the most commonly occurring combinations) and any other combination must be converted to an Escape sequence. The problem is that there are many more possible combinations of (last, run, level) than there are individual VLCs. 'Run' may take any value between 0 and 62; 'Level' any value between 1 and 128; and 'Last' is 0 or 1. This gives 16 002 possible combinations of (last, run, level). Three possible approaches to finding the VLC are as follows:

1.  Large look-up table indexed by (last, run, level). The size of this table may be reduced somewhat because only levels in the range 1–12 and runs in the range 0–40 have individual VLCs. The look-up procedure is as follows:

```
if ( | level | < 13 and run < 39)
    look up table based on (last, run, level)
    return individual VLC or calculate Escape sequence
else
    calculate Escape sequence
```

The look-up table has $(2 \times 40 \times 12) = 960$ entries; 102 of these contain individual VLCs and the remaining 858 contain a flag indicating that an Escape sequence is required.

2.  Partitioned look-up tables indexed by (last, run, level). Based on the values of last, run and level, choose a smaller look-up table (e.g. a table that only applies when last $= 0$). This requires one or more comparisons before choosing the table but allows the large table to be split into a number of smaller tables with fewer entries overall. The procedure is as follows:

```
if (last, run, level) ∈ {set A}
    look up table A
    return VLC or calculate Escape sequence
else if (last, run, level) ∈ {set B}
    look up table B
    return VLC or calculate Escape sequence
....
else
    calculate Escape sequence
```

For example, earlier versions of the H.263 'test model' software used this approach to reduce the number of entries in the partitioned look-up tables to 200 (i.e. 102 valid VLCs and 98 'empty' entries).

3. Conditional expression for every valid combination of (last, run, level). For example:

```
switch (last, run, level)
   case {A} : vlc = v_A, length = l_A
   case {B} : vlc = v_B, length = l_B
   ... (100 more cases) ...
   default : calculate Escape sequence
```

Comparing the three methods, method 1 lends itself to compact code, is easy to modify (by changing the look-up table contents) and is likely to be computationally efficient; however, it requires a large look-up table, most of which is redundant. Method 3, at the other extreme, requires the most code and is the most difficult to change (since each valid combination is 'hand-coded') but requires the least data storage. On some platforms it may be the slowest method. Method 2 offers a compromise between the other two methods.

*Hardware design*

A hardware architecture for variable length encoding performs similar tasks to those described above and an example is shown in Figure 8.11 (based on a design proposed by Lei and Sun[2]). A 'look-up' unit finds the length and value of the appropriate VLC and passes these to a 'pack' unit. The pack unit collects together a fixed number of bits (e.g. 8, 16 or 32 bits) and shifts these out to a stream buffer. Within the 'pack' unit, a counter records the number of bits in the output register. When this counter overflows, a data word is output (as in the example above) and the remaining upper bits in the output register are shifted down.

The design of the look-up unit is critical to the size, efficiency and adaptability of the design. Options range from a ROM or RAM-based look-up table containing all valid codes plus 'dummy' entries indicating that an Escape sequence is required, to a 'Hard-wired' approach (similar to the 'switch' statement described above) in which each valid combination is mapped to the appropriate VLC and length fields. This approach is sometimes described as a 'programmable logic array' (PLA) look-up table. Another example of a hardware VLE is presented elsewhere.[3]

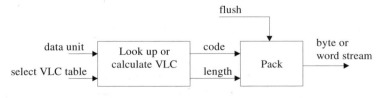

**Figure 8.11**   Hardware VLE

### 8.3.6   Variable Length Decoder Design

*Software design*

The operation of a decoder for VLCs can be summarised as follows:

```
scan through bits in an input buffer
if a valid VLC (length L) is detected
   remove L bits from buffer
   return corresponding data unit
if invalid VLC is detected
   return an error flag
```

Perhaps the most straightforward way of finding a valid VLC is to step through the relevant Huffman code tree. For example, a H.263 / MPEG-4 TCOEF code may be decoded by stepping through the tree shown in Figure 8.7, starting from the left:

```
if (first bit = 1)
   if (second bit = 1)
      if (third bit = 1)
         if (fourth bit = 1)
            return (0,0,2)
         else
            return (0,2,1)
      else
         return (0,1,1)
   else
      return (0,0,1)
else
   ... decode all VLCs starting with 0
```

This approach requires a large nested if...else statement (or equivalent) that can deal with 104 cases (102 unique TCOEF VLCs, one escape code, plus an error condition). This method leads to a large code size, may be slow to execute and is difficult to modify (because the Huffman tree is 'hand-coded' into the software); however, no extra look-up tables are required.

An alternative is to use one or more look-up tables. The maximum length of TCOEF VLC (excluding the sign bit and escape sequences) is 13 bits. We can construct a look-up table whose index is a 13-bit number (the 13 lsbs of the input stream). Each entry of the table contains either a (last, run, level) triplet or a flag indicating Escape or Error; $2^{13} = 8192$ entries are required, most of which will be duplicates of other entries. For example, every code beginning with '10...' (starting with the lsb) decodes to the triplet (0, 0, 1).

An initial test of the range of the 13-bit number may be used to select one of a number of smaller look-up tables. For example, the H.263 reference model decoder described earlier breaks the table into three smaller tables containing around 300 entries (about 200 of which are duplicate entries).

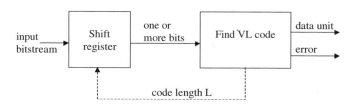

**Figure 8.12** Hardware VLD

The choice of algorithm may depend on the capabilities of the software platform. If memory is plentiful and array access relatively fast, a large look-up table may be the best approach for speed and flexibility. If memory is limited and/or array access is slow, better performance may be achieved with an 'if...else' approach or a partitioned look-up table.

Whichever approach is chosen, VL decoding requires a significant amount of bit-level processing and for many processors this makes it a computationally expensive function. An interesting development in recent years has been the emergence of dedicated hardware assistance for software VL decoding. The Philips TriMedia and Equator/Hitachi MAP platforms, for example, contain dedicated variable length decoder (VLD) co-processors that automatically decode VL data in an input buffer, relieving the main processor of the burden of variable length decoding.

*Hardware design*

Hardware designs for variable length decoding fall into two categories: (a) those that decode $n$ bits from the input stream every $m$ cycles (e.g. decoding 1 or 2 bits per cycle) and (b) those that decode $n$ complete VL codewords every $m$ cycles (e.g. decoding 1 codeword in one or two cycles). The basic architecture of a decoder is shown in Figure 8.12 (the dotted line 'code length L' is only required for category (b) decoders).

**Category (a), $n$ bits per $m$ cycles**   This type of decoder follows through the Huffman decoding tree. The simplest design processes one level of the tree every cycle: this is analogous to the large 'if...else' statement described above. The shift register shown in Figure 8.12 shifts 1 bit per cycle to the 'Find VL code' unit. This unit steps through the tree (based on the value of each input bit) until a valid code (a 'leaf') is found and can be implemented with a finite state machine (FSM) architecture. For example, Table 8.8 lists part of the FSM for the TCOEF tree shown in Figure 8.7. Each state corresponds to a node of the Huffman tree and the nodes in the table are labelled (with circles) in Figure 8.13 for convenience.

There are 102 nodes (and hence 102 states in the FSM) and 103 output values. To decode 1110, for example, the decoder traces the following sequence:

State 0 → State 2 → State 5 → State 6 → output (0, 2, 1)

Hence the decoder processes 1 bit per cycle (assuming that a state transition occurs per clock cycle).

**Table 8.8**  Part of state table for TCOEF decoding

| State | Input | Next state or output |
|-------|-------|----------------------|
| 0     | 0     | 1                    |
|       | 1     | 2                    |
| 1     | 0     | 3                    |
|       | 1     | 4                    |
| 2     | 0     | (0, 0, 1)            |
|       | 1     | 5                    |
| 3     | 0     | ... later states     |
|       | 1     | ...                  |
| 4     | 0     | ... later states     |
|       | 1     | ...                  |
| 5     | 0     | (0, 1, 1)            |
|       | 1     | 6                    |
| 6     | 0     | (0, 2, 1)            |
|       | 1     | (0, 0, 2)            |
| ...   | ...   | ...                  |

This type of decoder has the disadvantage that the processing rate depends on the (variable) rate of the coded stream. It is often more useful to be capable of processing one or more complete VLCs per clock cycle (for example, to guarantee a certain codeword throughput), and this leads to the second category of decoder design.

**Category (b), $n$ codewords per $m$ cycles**   This is analogous to the 'large look-up table' approach in a software decoder. $K$ bits (stored in the input shift register) are examined per cycle, where $K$ is the largest possible VLC size (13, excluding the sign bit, in the example of H.263/MPEG-4 TCOEF). The 'Find VL code' unit in Figure 8.12 checks all combinations of $K$ bits and finds a matching valid code, Escape code or flags an error. The length of the matching code ($L$ bits) is fed back and the shift register shifts the input data by $L$ bits (i.e. $L$ bits are removed from the input buffer). Hence a complete $L$-bit codeword can be processed in one cycle.

The shift register can be implemented using a barrel shifter (a shift-register circuit that shifts its contents by $L$ places in one cycle). The 'Find VL code' unit may be implemented using logic (a PLA). The logic array should minimise effectively since most of the possible input combinations are 'don't cares'. In the TCOEF example, all 13-bit input words '10XXXXXXXXXXX' map to the output (0, 0, 1). It is also possible to implement this unit as a ROM or RAM look-up table with $2^{13}$ entries.

A decoder that decodes one codeword per cycle is described by Lei and Sun[2] and Chang and Messerschmitt[4] examine the principles of concurrent VLC decoding. Further examples of VL decoders can be found elsewhere.[5,6]

## 8.3.7   Dealing with Errors

An error during transmission may cause the decoder to lose synchronisation with the sequence of VLCs and this in turn can cause incorrect decoding of subsequent VLCs. These

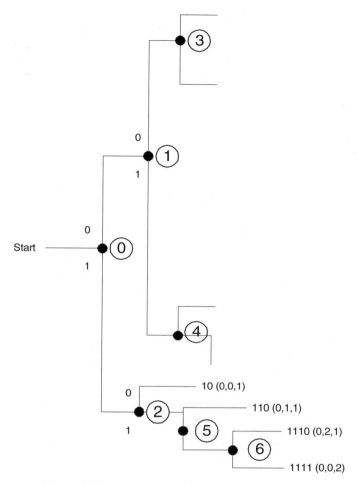

**Figure 8.13** Part of TCOEF tree showing state labels

decoding errors may continue to occur (propagate) until a resynchronisation point occurs in the bit stream. The synchronisation markers described in Section 8.2.2 limit the propagation of errors at the decoder. Increasing the frequency of synchronisation markers in the bit stream can reduce the effect of an error on the decoded image: however, markers are 'redundant' overhead and so this also reduces compression efficiency. Transmission errors and their effect on coded video are discussed further in Chapter 11.

Error-resilient alternatives to modified Huffman codes have been proposed. For example, MPEG-4 (video) includes an option to use reversible variable length codes (RVLCs), a class of codewords that may be successfully decoded in either a forward or backward direction from a resynchronisation point. When an error occurs, it is usually detectable by the decoder (since a serious decoder error is likely to violate the encoding syntax). The decoder can decode the current section of data in both directions, forward from the previous synchronisation point and backward from the next synchronisation point. Figure 8.14 shows an example. Region (a) is decoded and then an error is identified. The decoder 'skips' to the

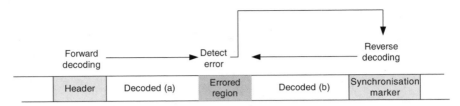

**Figure 8.14**   Decoding with RVLCs when an error is detected

next resynchronisation point and decodes backwards from there to recover region (b). Without RVLCs, all of region (b) would be lost.

An interesting recent development is the use of 'soft-decision' decoding of VLCs, utilising information available from the communications receiver about the probability of error in each codeword to improve decoding performance in the presence of channel noise.[7–9]

## 8.4   ARITHMETIC CODING

Entropy coding schemes based on codewords that are an integral number of bits long (such as Huffman coding or UVLCs) cannot achieve optimal compression of every set of data. This is because the theoretical optimum number of bits to represent a data symbol is usually a fraction (rather than an integer). This optimum number of bits is the 'information content' $\log_2(1/P)$, where $P$ is the probability of occurrence of each data symbol. In Table 8.1, for example, the motion vector '0.5' should be represented with 3.31 bits for maximum compression. Huffman coding produces a 5-bit codeword for this motion vector and so the compressed bit stream is likely to be larger than the theoretical maximally compressed bit stream.

Arithmetic coding provides a practical alternative to Huffman coding and can more closely approach the theoretical maximum compression.[10] An arithmetic encoder converts a sequence of data symbols into a single fractional number. The longer the sequence of symbols, the greater the precision required to represent the fractional number.

*Example*

Table 8.9 lists five motion vector values $(-2, -1, 0, 1, 2)$. The probability of occurrence of each vector is listed in the second column. Each vector is assigned a *subrange* within the

**Table 8.9**   Subranges

| Vector | Probability | $\log_2(1/P)$ | Subrange |
|--------|-------------|---------------|----------|
| −2 | 0.1 | 3.32 | 0–0.1 |
| −1 | 0.2 | 2.32 | 0.1–0.3 |
| 0 | 0.4 | 1.32 | 0.3–0.7 |
| 1 | 0.2 | 2.32 | 0.7–0.9 |
| 2 | 0.1 | 3.32 | 0.9–1.0 |

range 0.0–1.0, depending on its probability of occurrence. In this example, $(-2)$ has a probability of 0.1 and is given the subrange 0–0.1 (i.e. the first 10% of the total range 0–1.0). $(-1)$ has a probability of 0.2 and is given the next 20% of the total range, i.e. the subrange 0.1–0.3. After assigning a subrange to each vector, the total range 0–1.0 has been 'divided' amongst the data symbols (the vectors) according to their probabilities. The subranges are illustrated in Figure 8.15.

The encoding procedure is presented below, alongside a worked example for the sequence of vectors: $(0, -1, 0, 2)$.

*Encoding procedure*

| Encoding procedure | Range (L → H) | Symbol | Subrange (L → H) | Notes |
|---|---|---|---|---|
| 1. Set the initial range | 0 → 1.0 | | | |
| 2. For the first data symbol, find the corresponding subrange (low to high). | | (0) | 0.3 → 0.7 | |
| 3. Set the new range (1) to this subrange | 0.3 → 0.7 | | | |
| 4. For the next data symbol, find the subrange L to H | | (−1) | 0.1 → 0.3 | This is the subrange within the interval 0–1 |
| 5. Set the new range (2) to this subrange within the previous range | 0.34 → 0.42 | | | 0.34 is 10% of the range; 0.42 is 30% of the range |
| 6. Find the next subrange | | (0) | 0.3 → 0.7 | |
| 7. Set the new range (3) within the previous range | 0.364 → 0.396 | | | 0.364 is 30% of the range; 0.396 is 70% of the range |
| 8. Find the next subrange | | (2) | 0.9 → 1.0 | |
| 9. Set the new range (4) within the previous range | 0.3928 → 0.396 | | | 0.3928 is 90% of the range; 0.396 is 100% of the range |

Each time a symbol is encoded, the range (L to H) becomes progressively smaller. At the end of the encoding process (four steps in this example), we are left with a final range (L to H). The entire sequence of data symbols can be fully represented by transmitting a fractional number that lies within this final range. In the example above, we could send any number in

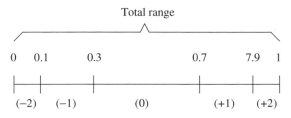

**Figure 8.15**  Subrange example

the range 0.3928–0.396: for example, 0.394. Figure 8.16 shows how the initial range (0–1) is progressively partitioned into smaller ranges as each data symbol is processed. After encoding the first symbol (vector 0), the new range is (0.3, 0.7). The next symbol (vector −1) selects the subrange (0.34, 0.42) which becomes the new range, and so on. The final symbol (vector +2) selects the subrange (0.3928, 0.396) and the number 0.394 (falling within this range) is transmitted. 0.394 can be represented as a fixed-point fractional number using 9 bits, i.e. our data sequence (0, − 1, 0, 2) is compressed to a 9-bit quantity.

## Decoding procedure

The sequence of subranges (and hence the sequence of data symbols) can be decoded from this number as follows.

| Decoding procedure | Range | Subrange | Decoded symbol |
|---|---|---|---|
| 1. Set the initial range | $0 \rightarrow 1$ | | |
| 2. Find the subrange in which the received number falls. This indicates the first data symbol | | $0.3 \rightarrow 0.7$ | (0) |
| 3. Set the new range (1) to this subrange | $0.3 \rightarrow 0.7$ | | |
| 4. Find the subrange *of the new range* in which the received number falls. This indicates the second data symbol | | $0.34 \rightarrow 0.42$ | (− 1) |
| 5. Set the new range (2) to this subrange within the previous range | $0.34 \rightarrow 0.42$ | | |
| 6. Find the subrange in which the received number falls and decode the third data symbol | | $0.364 \rightarrow 0.396$ | (0) |
| 7. Set the new range (3) to this subrange within the previous range | $0.364 \rightarrow 0.396$ | | |
| 8. Find the subrange in which the received number falls and decode the fourth data symbol | | $0.3928 \rightarrow 0.396$ | (2) |

The principal advantage of arithmetic coding is that the transmitted number (0.394 in this case, which can be represented as a fixed-point number with sufficient accuracy using 9 bits) is not constrained to an integral number of bits for each transmitted data symbol. To achieve optimal compression, the sequence of data symbols should be represented with:

$$\log_2(1/P_0) + \log_2(1/P_{-1}) - \log_2(1/P_0) + \log_2(1/P_2) \text{ bits } = 8.28 \text{ bits}$$

In this example, arithmetic coding achieves 9 bits which is close to optimum. A scheme using an integral number of bits for each data symbol (such as Huffman coding) would not come so close to the optimum number of bits and in general, arithmetic coding can outperform Huffman coding.

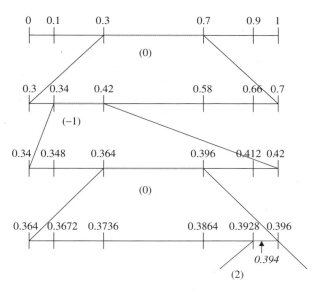

**Figure 8.16**   Arithmetic coding example

### 8.4.1   Implementation Issues

A number of practical issues need to be taken into account when implementing arithmetic coding in software or hardware.

*Probability distributions*

As with Huffman coding, it is not always practical to calculate symbol probabilities prior to coding. In several video coding standards (e.g. H.263, MPEG-4, H.26L), arithmetic coding is provided as an optional alternative to Huffman coding and pre-calculated subranges are   defined by the standard (based on 'typical' probability distributions). This has the advantage of avoiding the need to calculate and transmit probability distributions, but the disadvantage that compression will be suboptimal for a video sequence that does not exactly follow the standard probability distributions.

*Termination*

In our example, we stopped decoding after four steps. However, there is nothing contained in the transmitted number (0.394) to indicate the number of symbols that must be decoded: it could equally be decoded as three symbols or five. The decoder must determine when to stop decoding by some other means. In the arithmetic coding option specified in H.263, for example, the decoder can determine the number of symbols to decode according to the syntax of the coded data. Decoding of transform coefficients in a block halts when an end-of-block code is detected. Fixed-length codes (such as picture start code) are included in the bit stream and these will 'force' the decoder to stop decoding (for example, if a transmission error has occurred).

## Fixed-point arithmetic

Floating-point binary arithmetic is generally less efficient than fixed-point arithmetic and some processors do not support floating-point arithmetic at all. An efficient implementation with fixed-point arithmetic can be achieved by specifying the subranges as fixed-precision binary numbers. For example, in H.263, each subrange is specified as an unsigned 14-bit integer (i.e. a total range of 0–16 383). The subranges for the differential quantisation parameter *DQUANT* are listed as an example:

| H.263 *DQUANT* value | Subrange |
|:---:|:---:|
| 2 | 0–4094 |
| 1 | 4095–8191 |
| − 1 | 8192–12 286 |
| − 2 | 12 287–16 383 |

## Incremental encoding

As more data symbols are encoded, the precision of the fractional number required to represent the sequence increases. It is possible for the number to exceed the precision of the processor after a relatively small number of data symbols and a practical arithmetic encoder must take steps to ensure that this does not occur. This can be achieved by incrementally encoding bits of the fractional number as they are identified by the encoder. In our example above, after step 3, the range is 0.364–0.396. We know that the final fractional number will begin with '0.3 . . .' and so we can send the most significant part (e.g. 0.3, or its binary equivalent) without prejudicing the remaining calculations. At the same time, the limits of the range are left-shifted to extend the range. In this way, the encoder incrementally sends the most significant bits of the fractional number whilst continually readjusting the boundaries of the range to avoid arithmetic overflow.

## Patent issues

A number of patents have been filed that cover aspects of arithmetic encoding (such as IBM's 'Q-coder' arithmetic coding algorithm[11]). It is not entirely clear whether the arithmetic coding algorithms specified in the image and video coding standards are covered by patents. Some developers of commercial video coding systems have avoided the use of arithmetic coding because of concerns about potential patent infringements, despite its potential compression advantages.

## 8.5 SUMMARY

An entropy coder maps a sequence of data elements to a compressed bit stream, removing statistical redundancy in the process. In a block transform-based video CODEC, the main data elements are transform coefficients (run–level coded to efficiently represent sequences of zero coefficients), motion vectors (which may be differentially coded) and header information. Optimum compression requires the probability distributions of the data to be analysed prior to coding; for practical reasons, video CODECs use standard pre-calculated look-up tables for entropy coding.

The two most popular entropy coding methods for video CODECs are 'modified' Huffman coding (in which each element is mapped to a separate VLC) and arithmetic coding (in which a series of elements are coded to form a fractional number). Huffman encoding may be carried using a series of table look-up operations; a Huffman decoder identifies each VLC and this is possible because the codes are designed such that no code forms the prefix of any other. Arithmetic coding is carried out by generating and encoding a fractional number to represent a series of data elements.

This concludes the discussion of the main internal functions of a video CODEC (motion estimation and compensation, transform coding and entropy coding). The performance of a CODEC in a practical video communication system can often be dramatically improved by filtering the source video ('pre-filtering') and/or the decoded video frames ('post-filtering').

# REFERENCES

1. D. A. Huffman, 'A method for the construction of minimum-redundancy codes', *Proceedings of the Institute of Electrical and Radio Engineers*, **40**(9), September 1952.
2. S. M. Lei and M-T. Sun, 'An entropy coding system for digital HDTV applications', *IEEE Trans. CSVT*, **1**(1), March 1991.
3. Hao-Chieh Chang, Liang-Gee Chen, Yung-Chi Chang and Sheng-Chieh Huang, 'A VLSI architecture design of VLC encoder for high data rate video/image coding', 1999 IEEE International Symposium on Circuits and Systems (ISCAS '99).
4. S. F. Chang and D. Messerschmitt, 'Designing high-throughput VLC decoder, Part I – concurrent VLSI architectures', *IEEE Trans. CSVT*, **2**(2), June 1992.
5. J. Jeon, S. Park and H. Park, 'A fast variable-length decoder using plane separation', *IEEE Trans. CSVT*, **10**(5), August 2000.
6. B-J. Shieh, Y-S. Lee and C-Y. Lee, 'A high throughput memory-based VLC decoder with codeword boundary prediction', *IEEE Trans. CSVT*, **10**(8), December 2000.
7. A. Kopansky and M. Bystrom, 'Sequential decoding of MPEG-4 coded bitstreams for error resilience', *Proc. Conf. on Information Sciences and Systems*, Baltimore, 1999.
8. J. Wen and J. Villasensor, 'Utilizing soft information in decoding of variable length codes', *Proc. IEEE Data Compression Conference*, Utah, 1999.
9. S. Kaiser and M. Bystrom, 'Soft decoding of variable-length codes', *Proc. IEEE International Communications Conference*, New Orleans, 2000.
10. I. Witten, R. Neal and J. Cleary, 'Arithmetic coding for data compression', *Communications of the ACM*, **30**(6), June 1987.
11. J. Mitchell and W. Pennebaker, 'Optimal hardware and software arithmetic coding procedures for the Q-coder', *IBM Journal of Research and Development*, **32**(6), November 1988.

# 9

# Pre- and Post-processing

## 9.1   INTRODUCTION

The visual quality at the output of a video coding system depends on the performance of the 'core' coding algorithms described in Chapters 6–8, but can also be significantly influenced by pre- and post-processing (dealt with in this chapter) and bit-rate control (covered in Chapter 10).

In a practical video application, the source image is often far from perfect. Imperfections such as camera noise (introduced during image capture), poor lighting and camera shake may all affect the original images. Figure 9.1 shows a typical example of an image captured by a low-cost 'webcam'. Imperfections such as camera noise can produce variation and high-frequency activity in otherwise static parts of the video scene. These variations are likely to produce an increased number of transform coefficients and can significantly increase the amount of transmitted data (hence reducing compression performance). The aim of *pre-filtering* is to reduce these input impairments and improve compression efficiency whilst retaining the important features of the original image.

Quantisation leads to discontinuities and distortions in the transform coefficients that in turn produce *artefacts* (distortion patterns) in the decoded video images. In general, higher compression ratios require 'coarser' quantisation and introduce more obvious distortion. These artefacts are closely related to the block-based structure of transform coding and it is possible to detect and compensate using *post-filtering*. A significant improvement in subjective quality can be achieved by using filters designed to remove coding artefacts, in particular blocking and ringing. The aim of post-filtering is to reduce coding artefacts whilst retaining visually important image features.

Figure 9.2 shows the locations of pre- and post-filters in a video CODEC. In this chapter we investigate the causes of input variations and decoder artefacts and we examine a number of methods for improving subjective quality and compression efficiency through the use of filters.

## 9.2   PRE-FILTERING

DCT-based compression algorithms can perform well for smooth, noise-free regions of images. A region with flat texture or a gradual variation in texture (like the face area of the image in Figure 9.3) produces a very small number of significant DCT coefficients and hence is compressed efficiently. However, to generate a 'clean' video image like Figure 9.3 requires good lighting, an expensive camera and a high-quality video capture system. For most applications, these requirements are impractical. A typical desktop video-conferencing scenario might involve a low-cost camera on top of the user's monitor, poor lighting and a

**Figure 9.1**   Typical image from a low-cost 'webcam'

'busy' background, and all of these factors can be detrimental to the quality of the final image. A typical source image for this type of application is shown in Figure 9.1. Further difficulties can be caused for motion video compression: for example, a hand-held camera or a motorised surveillance camera are susceptible to camera shake which can significantly reduce the efficiency of motion estimation and compensation.

## 9.2.1   Camera Noise

Low-level noise with a uniform distribution is added to Figure 9.3 to produce Figure 9.4. The image now contains high-frequency variation which is not obvious but which will affect compression performance. After applying a DCT, this variation produces a number of high-frequency 'AC' coefficients, some of which remain after quantisation. This means that more bits will remain after compressing Figure 9.4 than after compressing the 'clean' image (Figure 9.3). After JPEG compression (with the same quantiser scale), Figure 9.3 compresses to 3211 bytes and Figure 9.4 compresses to 4063 bytes. The noise added to Figure 9.4 has decreased compression efficiency by over 25% in this example. This is typical of the effect produced by camera noise (i.e. noise introduced by the camera and/or analogue to

**Figure 9.2**   Pre- and post-filters in a video CODEC

**Figure 9.3**   Noise-free source image

digital conversion). All cameras/capture systems introduce noise, but it is more of a problem for lower-cost cameras (such as 'webcams').

Pre-filtering the image data before encoding can improve compression efficiency. The aim of a pre-filter is to increase compression performance without adversely affecting image quality, and a simple filter example is illustrated by Figure 9.5. The 'noisy' image (Figure 9.4) is filtered with a Gaussian 2-D spatial filter to produce Figure 9.5. This simple low-pass filter successfully reduces the noise. After JPEG compression, Figure 9.5 requires 3559 bytes (i.e. the compression efficiency is only 10% worse than the noise-free image). However, this compression gain is at the expense of a loss of image quality: the filter has 'blurred' some of the sharp lines in the image because it does not discriminate between high-frequency noise and 'genuine' high-frequency components of the image. With a more sophisticated pre-filter it is possible to minimise the noise whilst retaining important image features.

**Figure 9.4**   Image with noise

**Figure 9.5**   'Noisy' image after Gaussian filtering

## 9.2.2   Camera Movement

Unwanted camera movements (camera 'shake' or 'jitter') are another cause of poor compression efficiency. Block-based motion estimation performs best when the camera is fixed in one position or when it undergoes smooth linear movement (pan or tilt). In the case of a hand-held camera, or a motorised pan/tilt operation (e.g. as a surveillance camera 'sweeps' over a scene), the image tends to experience random 'jitter' between successive frames. If the motion search algorithm does not detect this jitter correctly, the result is a large residual frame after motion compensation. This in turn leads to a larger number of bits in the coded bit stream and hence poorer compression efficiency.

### *Example*

Two versions of a short 10-frame video sequence (the first frame is shown in Figure 9.6) are encoded with MPEG-4 (simple profile, with half-pixel motion estimation and

**Figure 9.6**   First frame of video sequence

compensation). Version 1 (the original sequence) has a fixed camera position. Version 2 is identical except that 2 of the 10 frames are shifted horizontally or vertically by up to 2 pixels (to simulate camera shake). The sequences are coded with H.263, using a fixed quantiser step size (10) in each case. For Version 1 (the original), the encoded sequence is 18 703 bits. For Version 2 (with 'shaking' of two frames), the encoded sequence is 29 080 bits: the compression efficiency drops by over 50% due to a small displacement within 2 of the 10 frames. This example shows that camera shake can be very detrimental to video compression performance (despite the fact that the encoder attempts to compensate for the motion).

The compression efficiency may be increased (and the subjective appearance of the video sequence improved) with automatic camera stabilisation. Mechanical stabilisation is used in some hand-held cameras, but this adds weight and bulk to the system. 'Electronic' image stabilisation can be achieved without extra hardware (at the expense of extra processing). For example, one method[1] attempts to stabilise the video frames prior to encoding. In this approach, a matching algorithm is used to detect global motion (i.e. common motion of all background areas, usually due to camera movement). The matching algorithm examines areas near the boundary of each image (not the centre of the image—since the centre usually contains foreground objects). If global motion is detected, the image is shifted to compensate for small, short-term movements due to camera shake.

## 9.3  POST-FILTERING

### 9.3.1  Image Distortion

Lossy image or video compression algorithms (e.g. JPEG, MPEG and H.26x) introduce distortion into video information. Higher compression ratios produce more distortion in the decoded video frames. The nature and appearance of the distortion depend on the type of compression algorithm. In a DCT-based CODEC, coding distortion is due to quantisation of DCT coefficients. This has two main effects on the DCT coefficients: those with smaller magnitudes (particularly higher-frequency AC coefficients) are set to zero, and the remaining coefficients (including the low-frequency and DC coefficients) lose precision due to quantisation. These effects lead to characteristic types of distortion in the decoded image. Figure 9.7 shows the result of encoding Figure 9.3 with baseline JPEG, at a compression ratio of 18.5× (i.e. the compressed image is 18.5 times smaller than the original). Figure 9.8 highlights three types of distortion in a close-up of this image, typical of any image or video sequence compressed using a DCT-based algorithm.

*Blocking*

Often, the most obvious distortion or artefact is the appearance of regular square blocks superimposed on the image. These *blocking artefacts* are a characteristic of block-based transform CODECs, and their edges are aligned with the $8 \times 8$ regions processed via the DCT. There are two causes of blocking artefacts: over-quantisation of the DC coefficient and suppression or over-quantisation of low frequency AC coefficients. The DC coefficient

**Figure 9.7**    Image compressed 18× (JPEG)

corresponds to the average (mean) value of each $8 \times 8$ block. In areas of smooth shading (such as the face area in Figure 9.7), over-quantisation of the DC coefficient means that there is a large change in level between neighbouring blocks. When two blocks with similar shades are quantized to different levels, the reconstructed blocks can have a larger 'jump' in level and hence a visible change of shade. This is most obvious at the block boundary, appearing as a tiling effect on smooth areas of the image. A second cause of blocking is over-quantisation or elimination of significant AC coefficients. Where there should be a smooth transition between blocks, a 'coarse' reconstruction of low-frequency basis patterns (see Chapter 7) leads to discontinuities between block edges. Figure 9.9 illustrates these two blocking effects in one dimension. Image sample amplitudes for a flat region are shown on the left and for a smoothly varying region on the right.

## Ringing

High quantisation can have a low-pass filtering effect, since higher-frequency AC coefficients tend to be removed during quantisation. Where there are strong edges in the original image, this low-pass effect can cause 'ringing' or 'ripples' near the edges. This is analogous to the effect of applying a low-pass filter to a signal with a sharp change in amplitude: low-frequency ringing components appear near the change position. This effect appears in Figure 9.8 as ripples near the edge of the hat.

## Basis pattern breakthrough

Coarse quantisation of AC coefficients can eliminate many of the original coefficients, leaving a few 'strong' AC coefficients in a block. After the inverse DCT, the basis pattern corresponding to a strong AC coefficient can appear in the reconstructed image block ('basis pattern breakthrough'). An example is highlighted in Figure 9.8: the block in question appears to be overlaid with one of the DCT basis patterns. Basis pattern breakthrough also

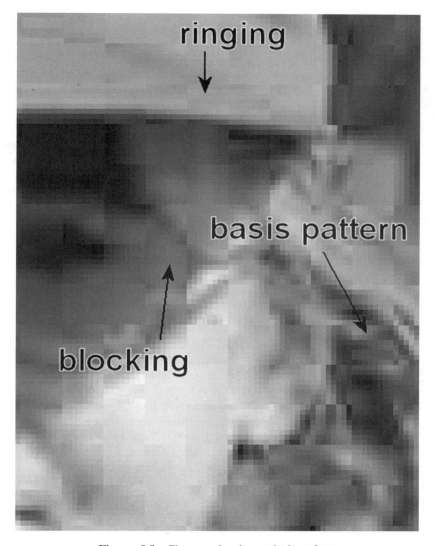

**Figure 9.8**  Close-up showing typical artefacts

contributes to the blocking effect (in this case, there is a sharp change between the strong basis pattern and the neighbouring blocks).

These three distortion effects degrade the appearance of decoded images or video frames. Blocking is particularly obvious because the large $8 \times 8$ patterns are clearly visible in highly compressed frames. The artefacts can also affect the performance of motion-compensated video coding. A video encoder that uses motion-compensated prediction forms a reconstructed (decoded) version of the current frame as a prediction reference for further encoded frames: this ensures that the encoder and decoder use identical reference frames and prevents 'drift' at the decoder. However, if a high quantiser scale is used, the reference frame at the encoder will contain distortion artefacts that were not present in the original frame. When

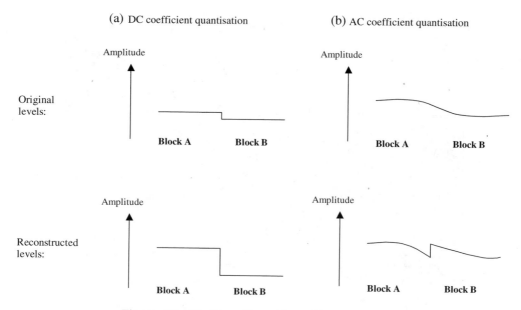

**Figure 9.9**   Blocking effects (shown in one dimension)

the reference frame (containing distortion) is subtracted from the next input frame (without distortion), these artefacts will tend to increase the energy in the motion-compensated residual frame, leading to a reduction in compression efficiency. This effect can produce a significant residual component even when there is no change between successive frames. Figure 9.10 illustrates this effect. The distorted reference frame (a) is subtracted from the current frame (b). There is no change in the image content but the difference frame (c) clearly contains residual energy (the 'speckled' effect). This residual energy will be encoded and transmitted, even though there is no real change in the image.

It is possible to design post-filters to reduce the effect of these predictable artefacts. The goal is to reduce the 'strength' of a particular type of artefact without adversely affecting the important features of the image (such as edges). Filters can be classified according to the type of artefact they are addressing (usually blocking or ringing), their computational complexity and whether they are applied inside or outside the coding 'loop'. A filter applied after decoding (outside the loop) can be made independent of the CODEC: however, good performance can be achieved by making use of parameters from the video decoder. A filter applied to the reconstructed frame within the encoder (inside the loop) has the advantage of improving compression efficiency (as described above) but must also be applied within the decoder. The use of in-loop filters is limited to non-standard CODECs except in the case of loop filters defined in the coding standards. Post-filters can be categorised as follows, depending on their position in the coding chain.

## (a) In-loop filters

The filter is applied to the reconstructed frame both in the encoder and in the decoder. Applying the filter within the encoder loop can improve the quality of the reconstructed

(a)

(b)

(c)

**Figure 9.10** (a) Reconstructed frame; (b) input frame; (c) difference frame

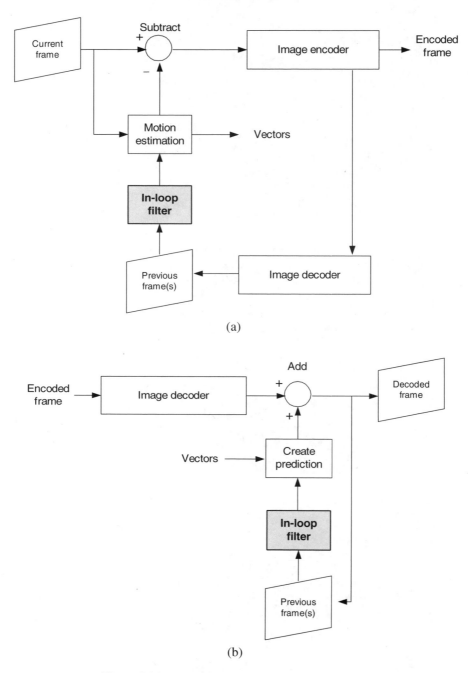

**Figure 9.11**   In-loop filter:  (a) encoder;  (b) decoder

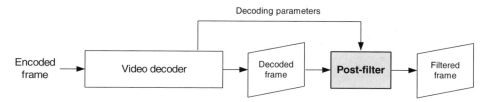

**Figure 9.12**  Decoder-dependent filter

reference frame, which in turn improves the accuracy of motion-compensated prediction for the *next* encoded frame since the quality of the prediction reference is improved. Figure 9.11 shows the position of the filter within the encoder and decoder, immediately prior to motion estimation or reconstruction. Placing the filter within the coding loop has two advantages: first, the decoder reference frame is identical to the encoder reference frame (avoiding prediction 'drift' between encoder and decoder) and second, the quality of the decoded frame is improved. The disadvantage of this approach is that the encoder and decoder must use an identical filter and this limits interoperability between CODECs (unless standardised filters are used, such as H.263 Annex J).

## (b) Decoder-dependent filters

In the second category, the filter is applied *after* the decoder and makes use of decoded parameters. A good example of a useful decoder parameter is the quantiser step size: this can be used to predict the expected level of distortion in the current block, e.g. more distortion is likely to occur when the quantiser step size is high than when it is low. This enables the decoder to adjust the 'strength' of the filter according to the expected distortion. A 'strong' filter may be applied when the quantiser step size is high, reducing the relevant type of distortion. A 'weak' filter is applied when the step size is low, preserving detail in blocks with lower distortion. Good performance can be achieved with this type of filter; however, the filter must be incorporated in the decoder or closely linked to decoding parameters. The location of the decoder-dependent filter is shown in Figure 9.12.

## (c) Decoder-independent filters

In order to minimise dependence on the decoder, the filter may be applied after decoding without any 'knowledge' of decoder parameters, as illustrated in Figure 9.13. This approach gives the maximum flexibility (for example, the decoder and the filter may be treated as separate 'black boxes' by the system designer). However, filter performance is generally not

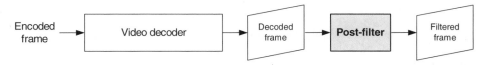

**Figure 9.13**  Decoder-independent filter

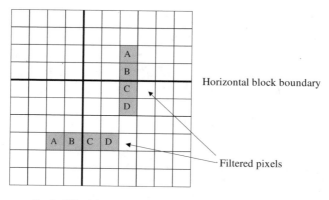

Vertical block boundary

**Figure 9.14**   H.263+ Annex J deblocking filter

as good as decoder-dependent filters, since the filter has no information about the coding of each block.

## 9.3.2   De-blocking Filters

Blocking artefacts are usually the most obvious and therefore the most important to minimise through filtering.

*In-loop filters*

It is possible to implement a non-standard in-loop de-blocking filter, however, the use of such a filter is limited to proprietary systems. Annex J of the H.263+ standard defines an optional de-blocking filter that operates within the encoder and decoder 'loops'. A 1-D filter is applied across block boundaries as shown in Figure 9.14. Four pixel positions at a time are smoothed across the block boundaries, first across the horizontal boundaries and then across the vertical boundaries. The 'strength' of the filter (i.e. the amount of smoothing applied to the pixels) is chosen depending on the quantiser value (as described above). The filter is effectively disabled if there is a strong discontinuity between the values of A and B or between the values of C and D: this helps to prevent filtering of 'genuine' strong horizontal or vertical edges in the original picture. In-loop de-blocking filters have been compared[2] and the authors conclude that the best performance is given by POCS algorithms (described briefly below).

*Decoder-dependent and decoder-independent filters*

If the filter is implemented only in the decoder (not in the encoder), the designer has complete flexibility and a wide range of filtering methods have been proposed.

Annex F of MPEG-4 describes an optional de-blocking filter that operates across each horizontal and vertical block boundary as above. The 'smoothness' of the image in the region

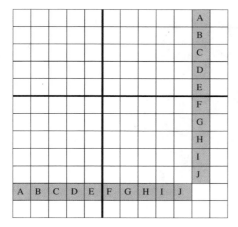

Horizontal block boundary

Vertical block boundary

**Figure 9.15**   MPEG-4 deblocking filter

of the boundary is estimated based on the values of 10 pixels (A to J in Figure 9.15). If the image is not deemed to be 'smooth', a default 1-D filter is applied to the two pixels on either side of the boundary (E and F). If the image is 'smooth' in this region, then a more sophisticated filter is required to reduce blocking whilst preserving the smooth texture: in this case, 8 pixels (B to I) are filtered. The filter parameters depend on the quantiser step size. By adapting the filter in this way, a more powerful (but more computationally complex) filter is applied where it is needed in smooth regions whilst a less complex filter is applied elsewhere.

Many alternative approaches can be found in the literature.[3–11] These range from highly complex algorithms such as Projection Onto Convex Sets (POCS), in which many candidate images are examined to find a close approximation to the decoded image that does not contain blocking artefacts, to algorithms such as the MPEG-4 Annex F filter that are significantly less complex. The best image quality is usually achieved at the expense of computation: for example, POCS algorithms are iterative and may be at least $20\times$ more complex than the decoding algorithm itself. Decoder-dependent algorithms can often outperform decoder-independent algorithms because the extra information about the coding parameters makes it easier to distinguish 'true' image features from blocking distortions.

### 9.3.3   De-ringing Filters

After blocking, ringing is often the next most obvious type of coding artefact. De-ringing filters receive somewhat less attention than de-blocking filters. MPEG-4 Annex F describes an optional post-decoder de-ringing filter. In this algorithm, a threshold *thr* is set for each reconstructed block based on the mean pixel value in the block. The pixel values within the block are compared with the threshold and $3 \times 3$ regions of pixels that are all either above or below the threshold are filtered using a 2-D spatial filter. This has the effect of smoothing homogeneous regions of pixels on either side of strong image edges whilst preserving the edges themselves: it is these regions that are likely to be affected by ringing. Figure 9.16

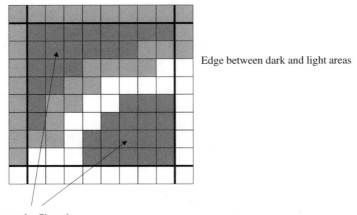

Edge between dark and light areas

Pixels that may be filtered

**Figure 9.16**   Application of MPEG-4 de-ringing filter

shows an example of regions of pixels that may be filtered in this way in a block containing a strong edge. In this example, pixels adjacent to the edge will be ignored by the filter (hence preserving the edge detail). Pixels in relatively 'flat' regions on either side of the edge (which are likely to contain ringing) will be filtered.

### 9.3.4   Error Concealment Filters

A final category of decoder filter is that of error concealment filters. When a decoder detects that a transmission error has occurred, it is possible to estimate the area of the frame that is likely to be corrupted by the error. Once the area is known, a spatial or temporal filter may be applied to attempt to conceal the error. Basic error concealment filters operate by interpolating from neighbouring error-free regions (spatially and/or temporally) to 'cover' the damaged area. More advanced methods (such as POCS filtering, mentioned above) attempt to maintain image features across the damaged region. Error concealment is discussed further in Chapter 11.

## 9.4   SUMMARY

Pre- and post-filtering can be valuable tools for a video CODEC designer. The goal of a pre-filter is to 'clean up' the source image and compensate for imperfections such as camera noise and camera shake whilst retaining visually important image features. A well-designed pre-filter can significantly improve compression efficiency by reducing the number of bits spent on coding noise. Post-filters are designed to compensate for characteristic artefacts introduced by block-based transform coding such as blocking and ringing effects. A post-filter can greatly improve subjective visual quality, reducing obvious distortions whilst retaining important features in the image. There are three main classes of this type of filter: loop filters (designed to improve motion compensation performance as well as image quality and present in both encoder and decoder), decoder-dependent post-filters (which make use of decoded parameters to improve filtering performance) and decoder-independent post-filters (which are independent of the coding algorithm but generally suffer from poorer performance than the other types). As with many other aspects of video CODEC design, there is usually a trade-off between filter complexity and performance (in terms of bit rate and image quality). The relationship between computational complexity, coded bit rate and image quality is discussed in the next chapter.

# REFERENCES

1. R. Kutka, 'Detection of image background movement as compensation for camera shaking with mobile platforms', *Proc. Picture Coding Symposium PCS01*, Seoul, April 2001.
2. M. Yuen and H. R. Wu, 'Performance comparison of loop filtering in generic MC/DPCM/DCT video coding', *Proc. SPIE Digital Video Compression*, San Jose, 1996.
3. Y. Yang, N. Galatsanos and A. Katsaggelos, 'Projection-based spatially adaptive reconstruction of block transform compressed images', *IEEE Trans. Image Processing*, **4**, July 1995.
4. Y. Yang and N. Galatsanos, 'Removal of compression artifacts using projections onto convex sets and line modeling', *IEEE Trans. Image Processing*, **6**, October 1997.
5. B. Jeon, J. Jeong and J. M. Jo, 'Blocking artifacts reduction in image coding based on minimum block boundary discontinuity', *Proc. SPIE VCIP95*, Taipei, 1995.
6. A. Nostratina, 'Embedded post-processing for enhancement of compressed images', *Proc. Data Compression Conference DCC-99*, Utah, 1999.
7. J. Chou, M. Crouse and K. Ramchandran, 'A simple algorithm for removing blocking artifacts in block transform coded images', *IEEE Signal Processing Letters*, **5**, February 1998.
8. S. Hong, Y. Chan and W. Siu, 'A practical real-time post-processing technique for block effect elimination', *Proc. IEEE ICIP96*, Lausanne, September 1996.
9. S. Marsi, R. Castagno and G. Ramponi, 'A simple algorithm for the reduction of blocking artifacts in images and its implementation', *IEEE Trans. on Consumer Electronics*, **44**(3), August 1998.
10. T. Meier, K. Ngan and G. Crebbin, 'Reduction of coding artifacts at low bit rates', *Proc. SPIE Visual Communications and Image Processing*, San Jose, January 1998.
11. Z. Xiong, M. Orchard, Y.-Q. Zhang, 'A deblocking algorithm for JPEG compressed images using overcomplete wavelet representations', *IEEE Trans. CSVT*, **7**, April 1997.

# 10

# Rate, Distortion and Complexity

## 10.1  INTRODUCTION

The choice of video coding algorithm and encoding parameters affect the coded bit rate and the quality of the decoded video sequence (as well as the computational complexity of the video CODEC). The precise relationship between coding parameters, bit rate and visual quality varies depending on the characteristics of the video sequence (e.g. 'noisy' input vs. 'clean' input; high detail vs. low detail; complex motion vs. simple motion). At the same time, practical limits determined by the processor and the transmission environment put constraints on the bit rate and image quality that may be achieved. It is important to control the video encoding process in order to maximise compression performance (i.e. high compression and/or good image quality) whilst remaining within the practical constraints of transmission and processing.

Rate-distortion optimisation attempts to maximise image quality subject to transmission bit rate constraints. The best optimisation performance comes at the expense of impractically high computation. Practical algorithms for the control of bit rate can be judged according to how closely they approach optimum performance. Many alternative rate control algorithms exist; sophisticated algorithms can achieve excellent rate-distortion performance, usually at a cost of increased computational complexity. The careful selection and implementation of a rate control algorithm can make a big difference to video CODEC performance.

Recent trends in software-only CODECs and video coding in power-limited environments (e.g. mobile computing) mean that computational complexity is an important factor in video CODEC performance. In many application scenarios, video quality is constrained by available computational resources as well as or instead of available bit rate. Recent developments in variable-complexity algorithms (VCAs) for video coding enable the developer to manage computational complexity and trade processing resources for image quality. This leads to situations in which rate, complexity and distortion are interdependent. New algorithms are required to jointly control bit rate and computational complexity whilst minimising distortion.

In this chapter we examine the factors that influence rate-distortion performance in a video CODEC and discuss how these factors can be exploited to efficiently control coded bit rate. We describe a number of popular algorithms for rate control. We discuss the relationship between computation, rate and distortion and show how new VCAs are beginning to influence the design of video CODECs.

## 10.2   BIT RATE AND DISTORTION

### 10.2.1   The Importance of Rate Control

A practical video CODEC operates within an environment that places certain constraints on its operation. One of the most important constraints is the rate at which the video encoder is 'allowed' to produce encoded data. A source of video data usually supplies video data at a constant bit rate (a constant number of bits per second) and a video encoder processes this high, constant-rate source to produce a compressed stream of bits at a reduced bit rate. The amount of compression (and hence the compressed bit rate) depends on a number of factors. These may include:

1. The encoding algorithm (intra-frame or inter-frame, forward or bidirectional prediction, integer or sub-pixel motion compensation, DCT or wavelet, etc.).

2. The type of video material (material containing lots of spatial detail and/or rapid movement generally produces more bits than material containing little detail and/or motion).

3. Encoding parameters and decisions (quantiser step size, picture or macroblock mode selection, motion vector search area, the number of intra-pictures, etc.).

Some examples of bit rate 'profiles' are given below. Figure 10.1 plots the number of bits in each frame for a video sequence encoded using Motion JPEG. Each frame is coded independently ('intra-coded') and the bit rate for each frame does not change significantly. Small variations in bit rate are due to changes in the spatial content of the frames in the 10-frame sequence. Figure 10.2 shows the bit rate variation for the same sequence coded

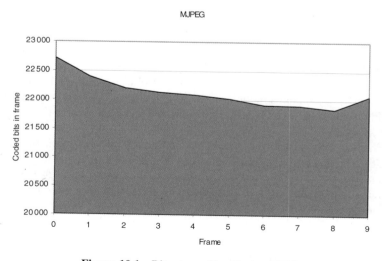

**Figure 10.1**   Bit-rate profile: Motion JPEG

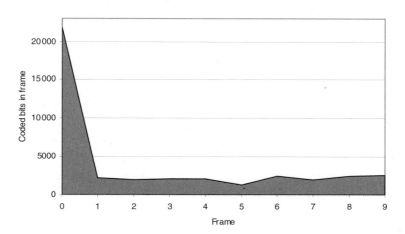

**Figure 10.2**   Bit-rate profile: H.263 (baseline)

with H.263. The first frame is an intra-frame and following frames are P-pictures. The compression efficiency for a P-picture is approximately 10 × higher than for an I-picture in this example and there is a small variation between P-pictures due to changes in detail and in movement. Coding the same sequence using MPEG-2 gives the bit rate profile shown in Figure 10.3. In this example, the initial I-picture is followed by the following sequence of picture types: B–B–P–B–B–P–B–B–I. There is clearly a large variation between the three picture types, with B-pictures giving the best compression performance. There is also a smaller variation between coded pictures of the same type (I, P or B) due to changes in detail and motion as before.

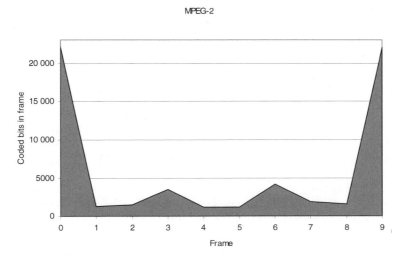

**Figure 10.3**   Bit-rate profile: MPEG-2

These examples show that the choice of algorithm and the content of the video sequence affect the bit rate (and also the visual quality) of the coded sequence. At the same time, the operating environment places important constraints on bit rate. These may include:

1. The mean bit rate that may be transmitted or stored.

2. The maximum bit rate that may be transmitted or stored.

3. The maximum variation in bit rate.

4. The requirement to avoid underflow or overflow of storage buffers within the system.

5. A requirement to minimise latency (delay).

*Examples:*

**DVD-video**    The mean bit rate is determined by the duration of the video material. For example, if a 3-hour movie is to be stored on a single 4.7 Gbyte DVD, then the mean bit rate (for the whole movie) must not exceed around 3.5 Mbps. The maximum bit rate is determined by the maximum transfer rate from the DVD and the throughput of the video decoder. Bit-rate variation (subject to these constraints) and latency are not such important issues.

**Video conferencing over ISDN**    The ISDN channel operates at a constant bit rate (e.g. 128 kbps). The encoded bit rate must match this channel rate exactly, i.e. no variation is allowed. The output of the video encoder is constant bit rate (CBR) coded video.

**Video conferencing over a packet-switched network**    The situation here is more complicated. The available mean and maximum bit rate may vary, depending on the network routeing and on the volume of other traffic. In some situations, latency and bit rate may be linked, i.e. a higher data rate may cause increased congestion and delay in the network. The video encoder can generate CBR or variable bit rate (VBR) coded video, but the mean and peak data rate may depend on the capacity of the network connection.

Each of these application examples has different requirements in terms of the rate of encoded video data. *Rate control*, the process of matching the encoder output to rate constraints, is a necessary component of the majority of practical video coding applications. The rate control 'problem' is defined below in Section 10.2.3. There are many different approaches to solving this problem and in a given situation, the choice of rate control method can significantly influence video quality at the decoder. Poor rate control may cause a number of problems such as low visual quality, fluctuations in visual quality and dropped frames leading to 'jerky' video.

   In the next section we will examine the relationship between coding parameters, bit rate and visual quality.

## 10.2.2   Rate-Distortion Performance

A lossless compression encoder produces a reduction in data rate with no loss of fidelity of the original data. A lossy encoder, on the other hand, reduces data rate at the expense of a loss of quality. As discussed previously, significantly higher compression of image and video data can be achieved using lossy methods than with lossless methods. The output of a lossy video CODEC is a sequence of images that are of a lower quality than the original images.

The *rate-distortion performance* of a video CODEC provides a measure of the image quality produced at a range of coded bit rates. For a given compressed bit rate, measure the distortion of the decoded sequence (relative to the original sequence). Repeat this for a range of compressed bit rates to obtain the *rate-distortion curve* such as the example shown in Figure 10.4. Each point on this graph is generated by encoding a video sequence using an MPEG-4 encoder with a different quantiser step size $Q$. Smaller values of $Q$ produce a higher encoded bit rate and lower distortion; larger values of $Q$ produce lower bit rates at the expense of higher distortion. In this figure, 'image distortion' is measured by peak signal to noise ratio (PSNR), described in Chapter 2. PSNR is a logarithmic measure, and a high value of PSNR indicates low distortion. The video sequence is a relatively static, 'head-and-shoulders' sequence ('Claire'). The shape of the rate-distortion curve is very typical: better image quality (as measured by PSNR) occurs at higher bit rates, and the quality drops sharply once the bit rate is below a certain threshold.

The rate-distortion performance of a video CODEC may be affected by many factors, including the following.

### *Video material*

Under identical encoding conditions, the rate-distortion performance may vary considerably depending on the video material that is encoded. Figure 10.5 compares the rate-distortion

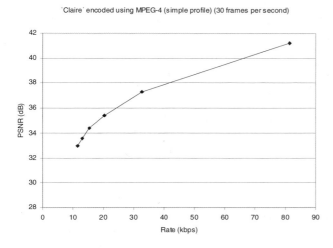

**Figure 10.4**   Rate-distortion curve example

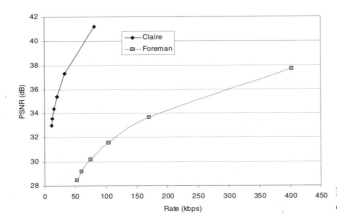

**Figure 10.5**  Rate-distortion comparison of two sequences

performance of two sequences, 'Claire' and 'Foreman', under identical encoding conditions (MPEG-4, fixed quantiser step size varying from 4 to 24). The 'Foreman' sequence contains a lot of movement and detail and is therefore more 'difficult' to compress than 'Claire'. At the same value of quantiser, 'Foreman' tends to have a much higher encoded bit rate and a higher distortion (lower PSNR) than 'Claire'. The shape of the rate-distortion curve is similar but the rate and distortion values are very different.

### Encoding parameters

In a DCT-based CODEC, a number of encoding parameters (in addition to quantiser step size) affect the encoded bit rate. An efficient motion estimation algorithm produces a small residual frame after motion compensation and hence a low coded bit rate; intra-coded macroblocks usually require more bits than inter-coded macroblocks; sub-pixel motion compensation produces a lower bit rate than integer–pixel compensation; and so on. Less obvious effects include, for example, the intervals at which the quantiser step size is varied during encoding. Each time the quantiser step size changes, the new value (or the change) must be signalled to the decoder and this takes more bits (and hence increases the coded bit rate).

### Encoding algorithms

Figures 10.1–10.3 illustrate how the coded bit rate changes depending on the compression algorithm. In each of these figures, the decoded image quality is roughly the same but there is a big difference in compressed bit rate.

### Rate control algorithms

A rate control algorithm chooses encoding parameters (such as those listed above) in order to try and achieve a 'target' bit rate. For a given bit rate, the choice of rate control algorithm can have a significant effect on rate-distortion performance, as discussed later in this chapter.

So far we have discussed only spatial distortion (the variation in quality of individual frames in the decoded video sequence). It is also important to consider temporal distortion, i.e. the situation where complete frames are 'dropped' from the original sequence in order to achieve acceptable performance. The curves shown in Figure 10.5 were generated for video sequences encoded at 30 frames per second. It would be possible to obtain lower *spatial* distortion by reducing the frame rate to 15 frames per second (dropping every second frame), at the expense of an increase in *temporal* distortion (because the frame rate has been reduced). The effect of this type of temporal distortion is apparent as 'jerky' video. This is usually just noticeable around 15–20 frames per second and very noticeable below 10 frames per second.

## 10.2.3   The Rate-Distortion Problem

The trade-off between coded bit rate and image distortion is an example of the general *rate-distortion problem* in communications engineering. In a lossy communication system, the challenge is to achieve a target data rate with minimal distortion of the transmitted signal (in this case, an image or sequence of images). This problem may be described as follows: 'Minimize distortion ($D$) whilst maintaining a bit rate $R$ that does not exceed a maximum bit rate $R_{max}$, or

$$\min\{D\} \text{ s.t. } R \leq R_{max} \tag{10.1}$$

(where s.t. means 'subject to').

The conditions of Equation 10.1 can be met by selecting the optimum encoding parameters to give the 'best' image quality (i.e. the lowest distortion) without exceeding the target bit rate. This process can be viewed as follows:

1. Encode a video sequence with a particular set of encoding parameters (quantiser step size, macroblock mode selection, etc.) and measure the coded bit rate and decoded image quality (or distortion). This gives a particular combination of rate ($R$) and distortion ($D$), an *R–D operating point*.

2. Repeat the encoding process with a different set of encoding parameters to obtain another *R–D* operating point.

3. Repeat for further combinations of encoding parameters. (Note that the set of possible combinations of parameters is very large.)

Figure 10.6 shows a typical set of operating points plotted on a graph. Each point represents the mean bit rate and distortion achieved for a particular set of encoding parameters. (Note that distortion [$D$] increases as rate [$R$] decreases). Figure 10.6 indicates that there are 'bad' and 'good' rate-distortion points. In this example, the operating points that give the best rate-distortion performance (i.e. the lowest distortion for a given rate $R$) lie close to the dotted curve. Rate-distortion theory tells us that this curve is convex (a *convex hull*). For a given

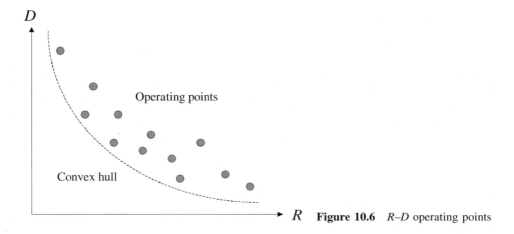

$R$   **Figure 10.6**   *R–D operating points*

target rate $R_{\text{max}}$, the minimum distortion $D$ occurs at a point on this convex curve. The aim of *rate-distortion optimisation* is to find a set of coding parameters that achieves an operating point as close as possible to this optimum curve.[1]

One way to find the position of the hull and hence achieve this optimal performance is by using *Lagrangian optimisation*. Equation 10.1 is difficult to minimise directly and a popular method is to express it in a slightly different way as follows:

$$\min\{J = D + \lambda R\} \tag{10.2}$$

$J$ is a new function that contains $D$ and $R$ (as before) as well as a *Lagrange multiplier*, $\lambda$. $J$ is the equation of a straight line $D + \lambda R$, where $\lambda$ gives the slope of the line. There is a solution to Equation 10.2 for every possible multiplier $\lambda$, and each solution is a straight line that makes a tangent to the convex hull described earlier. The procedure may be summarised as follows:

1. Encode the sequence many times, each time with a different set of coding parameters.

2. Measure the coded bit rate ($R$) and distortion ($D$) of each coded sequence. These measurements are the 'operating points' ($R$, $D$).

3. For each value of $\lambda$, find the operating point ($R$, $D$) that gives the smallest value $J$, where $J = D + \lambda R$. This gives one point on the convex hull.

4. Repeat step (3) for a range of $\lambda$ to find the 'shape' of the convex hull.

This procedure is illustrated in Figure 10.7. The ($R$, $D$) operating points are plotted as before. Three values of $\lambda$ are shown: $\lambda_1$, $\lambda_2$. and $\lambda_3$. In each case, the solution to $J = D + \lambda R$ is a straight line with slope $\lambda$. The operating point ($R$, $D$) that gives the smallest $J$ is shown in black, and these points occur on the lower boundary (the convex hull) of all the operating points.

The Lagrangian method will find the set (or sets) of encoding parameters that give the best performance and these parameters may then be applied to the video encoder to achieve optimum rate-distortion performance. However, this is usually a prohibitively complex

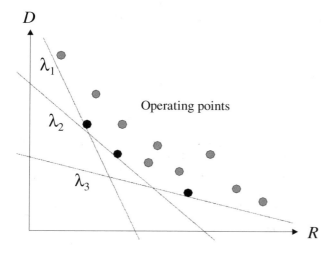

**Figure 10.7** Finding the best $(R, D)$ points using Lagrangian optimisation

process. Encoding decisions (such as quantiser step size, macroblock type, etc.) may change for every macroblock in the coded sequence and so there are an extremely large number of combinations of encoding parameters.

### Example

Macroblock 0 in a picture is encoded using MPEG-4 (simple profile) with a quantiser step size $Q_0$ in the range 2–31. The choice of $Q_1$ for macroblock 1 is constrained to $Q_0 +/- 2$. There are 30 possible values of $Q_0$; (almost) $30 \times 5 = 150$ possible combinations of $Q_0$ and $Q_1$; (almost) $30 \times 5 \times 5 = 750$ combinations of $Q_0$, $Q_1$ and $Q_2$; and so on.

The computation required to evaluate all possible choices of encoding decision becomes prohibitive even for a short video sequence. Furthermore, no two video sequences produce the same rate-distortion performance for the same encoding parameters and so this process needs to be carried out each time a sequence is to be encoded.

There have been a number of attempts to simplify the Lagrangian optimisation method in order to make it more practically useful.[2–4] For example, certain assumptions may be made about good and bad choices of encoding parameters in order to limit the exponential growth of complexity described above. The computational complexity of some of these methods is still much higher than the computation required for the encoding process itself: however, this complexity may be justified in some applications, such as (for example) encoding a feature film to obtain optimum rate-distortion performance for storage on a DVD.

An alternative approach is to estimate the optimum operating points using a model of the rate-distortion characteristic.[5] Lagrange-based optimisation is first carried out on some representative video sequences in order to find the 'true' optimal parameters for these sequences. The authors propose a simple model of the relationship between encoding mode selection and $\lambda$ and the encoding mode decisions required to achieve minimal distortion for a given rate constraint $R_{\mathrm{max}}$ can be estimated from this model. The authors report a clear performance gain over previous methods with minimal computational complexity. Another attempt has been made[6] to define an optimum partition between the coded bits representing motion vector information and the coded bits representing displaced frame difference (DFD) in an inter-frame CODEC.

### 10.2.4   Practical Rate Control Methods

Bit-rate control in a real-time video CODEC requires a relatively low-complexity algorithm. The choice of rate control algorithm can have a significant effect on video quality and many alternative algorithms have been developed. The choice of rate control algorithm is not straightforward because a number of factors are involved, including:

- the computational complexity of the algorithm
- whether the rate control 'model' is appropriate to the type of video material to be encoded (e.g. 'static' video-conferencing scenes or fast-action movies)
- the constraints of the transmission channel (e.g. low-delay real-time communications or offline storage).

A selection of algorithms is summarised here.

*Output buffer feedback*

One of the simplest rate control mechanisms is shown in Figure 10.8. A frame of video $i$ is encoded to produce $b_i$ bits. Because of the variation in content of a video sequence, $b_i$ is likely to vary from frame to frame, i.e. the encoder output bit rate is variable, $R_v$. In Figure 10.8 we assume that the channel rate is constant, $R_c$ (this is the case for many practical channels). In order to match the variable rate $R_v$ to the constant channel rate $R_c$, the encoded bits are placed in a buffer, filled at rate $R_v$ and emptied at rate $R_c$.

Figure 10.9 shows how the buffer contents vary during encoding of a typical video sequence. As each frame is encoded, the buffer fills at a variable rate and after encoding of each frame, a fixed number of bits $b_c$ are removed from the buffer. With no constraint on the

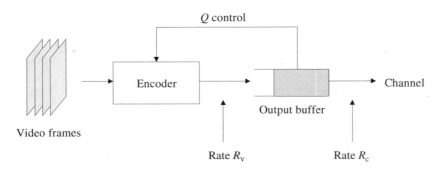

**Figure 10.8**   Buffer feedback rate control

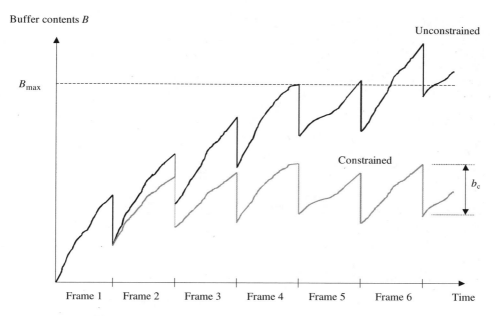

**Figure 10.9**   Buffer contents: constrained and unconstrained

variable rate $R_v$, it is possible for the buffer contents to rise to a point at which the buffer overflows ($B_{max}$ in the figure). The black line shows the unconstrained case: the buffer overflows in frames 5 and 6. To avoid this happening, a feedback constraint is required, where the buffer occupancy $B$ is 'fed back' to control the quantiser step size $Q$. As $B$ increases, $Q$ also increases which has the effect of increasing compression and reducing the number of bits per frame $b_i$. The grey line in Figure 10.9 shows that with feedback, the buffer contents are never allowed to rise above about 50% of $B_{max}$.

This method is simple and straightforward but has several disadvantages. A sudden increase in activity in the video scene may cause $B$ to increase too rapidly to be effectively controlled by the quantiser $Q$, so that the buffer overflows, and in this case the only course of action is to skip frames, resulting in a variable frame rate. As Figure 10.9 shows, $B$ increases towards the end of each encoded frame and this means that $Q$ also tends to increase towards the end of the frame. This can lead to an effect whereby the top of each frame is encoded with a relatively high quality whereas the foot of the frame is highly quantised and has an obvious drop in quality, as shown in Figure 10.10. The basic buffer-feedback method tends to produce decoded video with obvious quality variations.

## MPEG-2 Test Model 5[7]

Version 5 of the MPEG-2 video Test Model (a reference design for MPEG-2 encoding and decoding) describes a rate control algorithm for CBR encoding that takes account of the different properties of the three coded picture types (I, P and B-pictures). The algorithm consists of three steps: bit allocation, rate control and modulation.

low quantiser

medium quantiser

high quantiser

**Figure 10.10**  Frame showing quality variation

1. Bit allocation:
   (a) assign a target number of bits to the current GOP (based on the target constant bit rate);
   (b) assign a target number of bits $T$ to the current picture based on:
       – the 'complexity' of the previous picture of the same type (I, P, B) (i.e. the level of temporal and/or spatial activity);
       – the target number of bits for the GOP.

2. Rate control:
   (a) during encoding of the current picture, maintain a count of the number of coded bits so far, $d$;
   (b) compare $d$ with the target total number of bits $T$ and choose the quantiser step size $Q$ to try and meet the target $T$.

3. Modulation:
   (a) measure the variance of the luminance data in the current macroblock;
   (b) if the variance is higher than average (i.e. there is a high level of detail in the current region of the picture), increase $Q$ (and hence increase compression).

The aim of this rate control algorithm is to:

- achieve a target number of coded bits for the current GOP;
- deal with I, P and B-pictures separately;
- quantise areas of high detail more 'coarsely'.

This last aim should give improved subjective visual quality since the human eye is more sensitive to coarse quantisation (high distortion) in areas of low detail (such as a smooth region of the picture).

## H.263 Test Model 8[8]

Version 8 (and later versions) of the H.263 Test Model use a rate control algorithm that consists of frame-level rate control (determines whether to skip or code the current frame) and macroblock-level rate control (calculate the quantisation step size for each macroblock).

**Frame level control**   Each encoded frame adds to the encoder output buffer contents; each transmitted frame removes bits from the output buffer. If the number of bits in the buffer exceeds a threshold $M$, skip the next frame; otherwise set a target number of bits $B$ for encoding the next frame. A higher threshold $M$ means fewer skipped frames, but a larger delay through the system.

**Macroblock level control**   This is based on a model for the number of bits $B_i$ required to encode macroblock $i$ (Equation 10.3):

$$B_i = A \left( K \frac{\sigma_i^2}{Q_i^2} + C \right) \tag{10.3}$$

$A$ is the number of pixels in a macroblock, $\sigma_i$ is the standard deviation of luminance and chrominance in the residual macroblock (i.e. a measure of variation within the macroblock), $Q_i$ is the quantisation step size and $K$ and $C$ are constant model parameters. The following steps are carried out for each macroblock $i$:

1. Measure $\sigma_i$.

2. Calculate $Q_i$ based on $B$, $K$, $C$, $\sigma_i$ and a macroblock weight $\alpha_i$.

3. Encode the macroblock.

4. Update the model parameters $K$ and $C$ based on the actual number of coded bits produced for the macroblock.

The weight $\alpha_i$ controls the 'importance' of macroblock $i$ to the subjective appearance of the image: a low value of $\alpha_i$ means that the current macroblock is likely to be highly quantised. In the test model, these weights are selected to minimise changes in $Q_i$ at lower bit rates because each change involves sending a modified quantisation parameter $DQUANT$ which means encoding an extra 5 bits per macroblock. It is important to minimise the number of changes to $Q_i$ during encoding of a frame at low bit rates because the extra 5 bits in a macroblock may become significant; at higher bit rates, this $DQUANT$ overhead is less important and we may change $Q$ more frequently without significant penalty.

This rate control method is effective at maintaining good visual quality with a small encoder output buffer which keeps coding delay to a minimum (important for low-delay real-time communications).

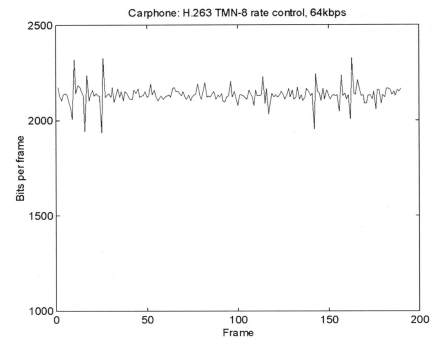

**Figure 10.11**   Bit-rate profile: H.263 TM8

*Example*

A 200-frame video sequence, 'Carphone', is encoded using H.263 with TM8 rate control. The original frame rate is 30 frames per second, QCIF resolution, and the target bit rate is 64 kbps. Figure 10.11 shows the bit-rate variation during encoding. In order to achieve 64 kbps without dropping any frames, the mean bit rate should be 2133 bits per frame, and the encoder clearly manages to maintain this bit rate (with occasional variations of about $+/-10\%$). Figure 10.12 shows the PSNR of each frame in the sequence after encoding and decoding. Towards the end of the sequence, the movement in the scene increases and it becomes 'harder' to code efficiently. The rate control algorithm compensates for this by increasing the quantiser step size and the PSNR drops accordingly. Out of the original 200 frames, the encoder has to drop 6 frames to avoid buffer overflow.

*MPEG-4 Annex L*

The MPEG-4 video standard describes an optional rate control algorithm in Annex L.1, known as the Scalable Rate Control (SRC) scheme. This algorithm is appropriate for a single video object (i.e. a rectangular VO that covers the entire frame) and a range of bit rates and spatial/temporal resolutions. The scheme described in Annex L offers rate control at the frame-level only (i.e. a single quantiser step size is chosen for a complete frame). The SRC attempts to achieve a target bit rate over a certain number of frames (a 'segment' of frames, usually starting with an I-picture).

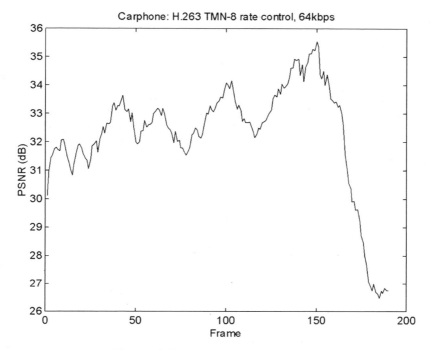

**Figure 10.12**   PSNR profile: H.263 TM8

The SRC scheme assumes the following model for the encoder rate $R$:

$$R = \frac{X_1 S}{Q} + \frac{X_2 S}{Q^2}$$   (10.4)

$Q$ is the quantiser step size, $S$ is the mean absolute difference of the residual frame after motion compensation and $X_1$, $X_2$ are model parameters. $S$ provides a measure of frame complexity (easier to compute than the standard deviation $\sigma$ used in the H.263 TM8 rate control scheme because the absolute difference, SAE, is calculated during motion estimation).

Rate control consists of the following steps which are carried out after motion compensation and before encoding of each frame $i$:

1. Calculate a target bit rate $R_i$, based on the number of frames in the segment, the number of bits that are available for the remainder of the segment, the maximum acceptable buffer contents and the estimated complexity of frame $i$. (The maximum buffer size affects the latency from encoder input to decoder output. If the previous frame was complex, it is assumed that the next frame will be complex and should therefore be allocated a suitable number of bits: the algorithm attempts to balance this requirement against the limit on the total number of bits for the segment.)

2. Compute the quantiser step size $Q_i$ (to be applied to the whole frame), calculate $S$ for the complete residual frame and solve Equation 10.4 to find $Q$.

3. Encode the frame.

4. Update the model parameters $X_1$, $X_2$ based on the actual number of bits generated for frame $i$.

The SRC algorithm differs from H.263 TM8 in two significant ways: it aims to achieve a target bit rate across a segment of frames (rather than a sequence of arbitrary length) and it does not modulate the quantiser step size within a coded frame (this can give a more uniform visual appearance within each frame but makes it difficult to maintain a small buffer size and hence a low delay). An extension to the SRC is described in Annex L.3 of MPEG-4 which supports modulation of the quantiser step size at the macroblock level and is therefore more suitable for low-delay applications. The macroblock rate control extension (L.3) is similar to H.263 Test Model 8 rate control.

The SRC algorithm is described in some detail in the MPEG-4 standard; a further discussion of MPEG-4 rate control issues can be found elsewhere.[9]

## 10.3    COMPUTATIONAL COMPLEXITY

### 10.3.1    Computational Complexity and Video Quality

So far we have considered the trade-off between bit rate and video quality. The discussion of rate distortion in Section 10.2.3 highlighted another trade-off between computational complexity and video quality. A video coding algorithm that gives excellent rate-distortion performance (good visual quality for a given bit rate) may be impractical because it requires too much computation.

There are a number of cases where it is possible to achieve higher visual quality at the expense of increased computation. A few examples are listed below:

- *DCT block size*: better decorrelation can be achieved with a larger DCT block size, at the expense of higher complexity. The $8 \times 8$ block size is popular because it achieves reasonable performance with manageable computational complexity.

- *Motion estimation search algorithm*: full-search motion estimation (where every possible match is examined within the search area) can outperform most reduced-complexity algorithms. However, algorithms such as the 'three step search' which sample only a few of the possible matches are widely used because they reduce complexity at the expense of a certain loss of performance.

- *Motion estimation search area*: a good match (and hence better rate-distortion performance) is more likely if the motion estimation search area is large. However, practical video encoders limit the search area to keep computation to manageable levels.

- *Rate-distortion optimisation*: obtaining optimal (or even near-optimal) rate-distortion performance requires computationally expensive optimisation of encoding parameters, i.e. the best visual quality for a given bit rate is achieved at the expense of high complexity.

- *Choice of frame rate*: encoding and decoding computation increases with frame rate and it may be necessary to accept a low frame rate (and 'jerky' video) because of computational constraints.

These examples show that many aspects of video encoding and decoding are a trade-off between computation and quality. Traditionally, hardware video CODECs have been designed with a fixed level of computational performance. The architecture and the clock rate determine the maximum video processing rate. Motion search area, block size and maximum frame rate are fixed by the design and place a predetermined 'ceiling' on the rate-distortion performance of the CODEC.

Recent trends in video CODEC design, however, require a more flexible approach to these trade-offs between complexity and quality. The following scenarios illustrate this.

## Scenario 1: Software video CODEC

Video is captured via a capture board or 'webcam'. Encoding, decoding and display are carried out entirely in software. The 'ceiling' on computational complexity depends on the available processing resources. These resources are likely to vary from platform to platform (for example, depending on the specification of a PC) and may also vary depending on the number of other applications contending for resources. Figure 10.13 compares the resources available to a software CODEC in two cases: when it is the only intensive application running, the CODEC has most of the system resources available, whereas when the CODEC must contend with other applications, fewer processing cycles are available to it. The computational resources (and therefore the maximum achievable video quality) are no longer fixed.

## Scenario 2: Power-limited video CODEC

In a mobile or hand-held computing platform, power consumption is at a premium. It is now common for a processor in a portable PC or personal digital assistant to be 'power-aware', e.g. a laptop PC may change the processor clock speed depending on whether it is running from a battery or from an AC supply. Power consumption increases depending on the activity of peripherals, e.g. hard disk accesses, display activity, etc. There is therefore a need to manage and limit computation in order to maximise battery life.

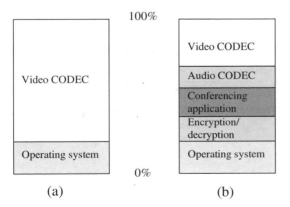

**Figure 10.13** Available computational resources

These scenarios illustrate the need for a more flexible approach to computation in a video CODEC. In this type of scenario, computation can no longer be considered to be a 'constant'. CODEC performance is now a function of three variables: computational complexity, coded bit rate and video quality. Optimising the complexity, rate and distortion performance of a video CODEC requires flexible control of computational complexity and this has led to the development of *variable complexity algorithms* for video coding.

## 10.3.2   Variable Complexity Algorithms

A variable complexity algorithm (VCA) carries out a particular task with a controllable degree of computational overhead. As discussed above, computation is often related to image quality and/or compression efficiency: in general, better image quality and/or higher compression require a higher computational overhead.

### Input-independent VCAs

In this class of algorithms, the computational complexity of the algorithm is independent of the input data. Examples of input-independent VCAs include:

- *Frame skipping*: encoding a frame takes a certain amount of processing resources and 'skipping' frames (i.e. not coding certain frames in the input sequence) is a crude but effective way of reducing processor utilisation. The relationship between frame rate and utilisation is not necessarily linear in an inter-frame CODEC: when the frame rate is low (because of frame skipping), there is likely to be a larger difference between successive frames and hence more data to code in the residual frame. Frame skipping may lead to a variable frame rate as the available resources change and this can be very distracting to the viewer. Frame skipping is widely used in software video CODECs.

- *Motion estimation (ME) search window*: increasing or decreasing the ME search window changes the computational overhead of motion estimation. The relationship between search window size and computational complexity depends on the search algorithm. Table 10.1 compares the overhead of different search window sizes for the popular $n$-step search algorithm. With no search, only the (0, 0) position is matched; with a search window of $+/-1$, a total of nine positions are matched; and so on.

**Table 10.1**   Computational overhead for $n$-step search (integer search)

| Search window | Number of comparison steps | Computation (normalised) |
|:---:|:---:|:---:|
| 0 | 1 | 0.03 |
| $+/-1$ | 9 | 0.27 |
| $+/-3$ | 17 | 0.51 |
| $+/-7$ | 25 | 0.76 |
| $+/-15$ | 33 | 1.0 |

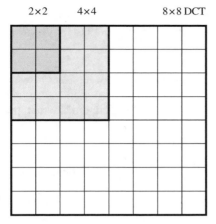

2×2     4×4          8×8 DCT

**Figure 10.14**   Pruned DCT

- *Pruned DCT*: a forward DCT (FDCT) processes a block of samples (typically $8 \times 8$) and produces a block of coefficients. In a typical image block, many of the coefficients are zero after quantisation and only a few non-zero coefficients remain to be coded and transmitted. These non-zero coefficients tend to occupy the lower-frequency positions in the block. A 'pruned' DCT algorithm only calculates a subset of the $8 \times 8$ DCT coefficients (usually lower frequencies), reducing the computational overhead of the DCT.[10,11] Examples of possible subsets are shown in Figure 10.14: the 'full' $8 \times 8$ DCT may be reduced to a $4 \times 4$ or $2 \times 2$ DCT, producing only low-frequency coefficients. However, applying a pruned DCT to all blocks means that the small (but significant) number of high-frequency coefficients are lost and this can have a very visible impact on image quality.

## Input-dependent algorithms

An input-dependent VCA controls computational complexity depending on the characteristics of the video sequence or coded data. Examples include the following.

**Zero testing in IDCT**   In a DCT-based CODEC operating at medium or low bit rates, many blocks contain no AC coefficients after quantisation (i.e. only the DC coefficient remains, or no coefficients remain). This may be exploited to reduce the complexity of the IDCT (which must be calculated in both the encoder and the decoder in an inter-frame CODEC). Each row or column of eight coefficients is tested for zeros. If the seven highest coefficients are all zero, then the row or column will contain a uniform value (the DC coefficient) after the IDCT. In this case, the IDCT may be skipped and all samples set to the DC value:

```
if (F1 = F2 = F3 = F4 = F5 = F6 = F7 = 0) {
   f0 = f1 = f2 = f3 = f4 = f5 = f6 = f7 = F0
} else {
   [calculate the IDCT..]
}
```

There is a small overhead associated with testing for zero: however, the computational saving can be very significant and there is no loss of quality. Further input-dependent complexity reductions can be applied to the IDCT.[12]

**FDCT complexity reduction**   Many blocks contain few non-zero coefficients after quantisation (particularly in inter-coded macroblocks). It is possible to predict the occurrence of some of these blocks before the FDCT is carried out so that the FDCT and quantisation steps may be skipped, saving computation. The sum of absolute differences (SAD or SAE) calculated during motion estimation can act as a useful predictor for these blocks. SAD is proportional to the energy remaining in the block after motion compensation. If SAD is low, the energy in the residual block is low and it is likely that the block will contain little or no data after FDCT and quantisation. Figure 10.15 plots the probability that a block contains no coefficients after FDCT and quantisation, against SAD. This implies that it should be possible to skip the FDCT and quantisation steps for blocks with an SAD of less than a threshold value $T$:

```
if (SAD < T) {
    set block contents to zero
} else {
    calculate the FDCT and quantize
}
```

If we set $T = 200$ then any block with SAD $< 200$ will not be coded. According to the figure, this 'prediction' of zero coefficients will be correct 90% of the time. Occasionally (10% of the time in this case), the prediction will fail, i.e. a block will be skipped that should have been encoded. The reduction in complexity due to skipping FDCT and quantisation for some blocks is therefore offset by an increase in distortion due to incorrectly skipped blocks.[13,14]

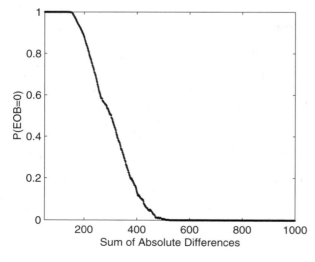

**Figure 10.15**  Probability of zero block vs. SAD

**Input-dependent motion estimation**    A description has been given[15] of a motion estimation algorithm with variable computational complexity. This is based on the nearest neighbours search (NNS) algorithm (described in Chapter 6), where motion search positions are examined in a series of 'layers' until a minimum is detected. The NNS algorithm is extended to a VCA by adding a computational constraint on the number of layers that are examined at each iteration of the algorithm. As with the SAD prediction discussed above, this algorithm reduces computational complexity at the expense of increased coding distortion. Other computationally scalable algorithms for motion estimation algrithms are described elsewhere.[16,17]

## 10.3.3    Complexity-Rate Control

The VCAs described above are useful for controlling the computational complexity of video encoding and decoding. Some VCAs (such as zero testing in the IDCT) have no effect on image quality; however, the more flexible and powerful VCAs (such as zero DCT prediction) do have an effect on quality. These VCAs may also change the coded bit rate: for example, if a high proportion of DCT operations are 'skipped', fewer coded bits will be produced and the rate will tend to drop. Conversely, the 'target' bit rate can affect computational complexity if VCAs are used. For example, a lower bit rate and higher quantiser scale will tend to produce fewer DCT coefficients and a higher proportion of zero blocks, reducing computational complexity.

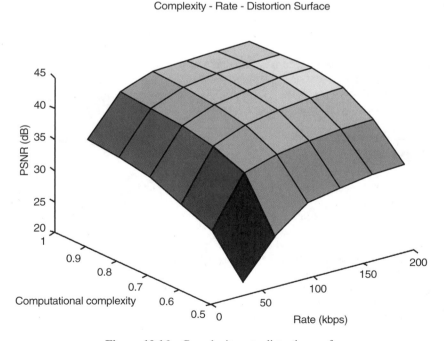

**Figure 10.16**    Complexity-rate-distortion surface

It is therefore not necessarily correct to treat complexity control and rate control as separate issues. An interesting recent development is the emergence of *complexity-distortion theory*.[18] Traditionally, video CODECs have been judged by their rate-distortion performance as described in Section 10.2.2. With the introduction of VCAs, it becomes necessary to examine performance in three axes: complexity, rate and distortion. The 'operating point' of a video CODEC is no longer restricted to a rate-distortion *curve* but instead lies on a rate-distortion-complexity *surface*, like the example shown in Figure 10.16. Each point on this surface represents a possible set of encoding parameters, leading to a particular set of values for coded bit rate, distortion and computational complexity.

Controlling rate involves moving the operating point along this surface in the rate-distortion plane; controlling complexity involves moving the operating point in the complexity-distortion plane. Because of the interrelationship between computational complexity and bit rate, it may be appropriate to control complexity and rate at the same time. This new area of complexity-rate control is at a very early stage and some preliminary results can be found elsewhere.[14]

## 10.4   SUMMARY

Many practical video CODECs have to operate in a rate-constrained environment. The problem of achieving the best possible rate-distortion performance is difficult to solve and optimum performance can only be obtained at the expense of prohibitively high computational cost. Practical rate control algorithms aim to achieve good, consistent video quality within the constraints of rate, delay and complexity. Recent developments in variable complexity coding algorithms enable a further trade-off between computational complexity and distortion and are likely to become important for CODECs with limited computational resources and/or power consumption.

Bit rate is one of a number of constraints that are imposed by the transmission or storage environment. Video CODECs are designed for use in communication systems and these constraints must be taken into account. In the next chapter we examine the key 'quality of service' parameters required by a video CODEC and provided by transmission channels.

## REFERENCES

1. A. Ortega and K. Ramchandran, 'Rate-distortion methods for image and video compression', *IEEE Signal Processing Magazine*, November 1998.
2. L-J Lin and A. Ortega, 'Bit-rate control using piecewise approximated rate-distortion character-istics', *IEEE Trans. CSVT*, **8**, August 1998.
3. Y. Yang, 'Rate control for video coding and transmission', Ph.D. Thesis, Cornell University, 2000.
4. M. Gallant and F. Kossentini, 'Efficient scalable DCT-based video coding at low bit rates', *Proc. ICIP99*, Japan, October, 1999.
5. G. Sullivan and T. Wiegand, 'Rate-distortion optimization for video compression', *IEEE Signal Processing Magazine*, November 1998.
6. G. M. Schuster and A. Katsaggelos, 'A theory for the optimal bit allocation between displacement vector field and displaced frame difference', *IEEE J. Selected Areas in Communications*, **15**(9), December 1997.

7. ISO/IEC JTC1/SC29 WG11 Document 93/457, 'MPEG-2 Video Test Model 5', Sydney, April 1993.

8. J. Ribas-Corbera and S. Lei, 'Rate control for low-delay video communications [H.263 TM8 rate control]', ITU-T Q6/SG16 Document Q15-A-20, June 1997.

9. J. Ronda, M. Eckert, F. Jaureguizar and N. Garcia, 'Rate control and bit allocation for MPEG-4', *IEEE Trans. CSVT*, **9**(8), December 1999.

10. C. Christopoulos, J. Bormans, J. Cornelis and A. N. Skodras, 'The vector-radix fast cosine transform: pruning and complexity analysis', *Signal Processing*, **43**, 1995.

11. A. Hossen and U. Heute, 'Fast approximate DCT: basic idea, error analysis, applications', *Proc. ICASSP97*, Munich, April 1997.

12. K. Lengwehasatit and A. Ortega, 'DCT computation based on variable complexity fast approximations', *Proc. ICIP98*, Chicago, October 1998.

13. M-T. Sun and I-M. Pao, 'Statistical computation of discrete cosine transform in video encoders', *J. Visual Communication and Image Representation*, June 1998.

14. I. E. G. Richardson and Y. Zhao, 'Video CODEC complexity management', *Proc. PCS01*, Seoul, April 2001.

15. M. Gallant, G. Côté and F. Kossentini, 'An efficient computation-constrained block-based motion estimation algorithm for low bit rate video coding', *IEEE Trans. Image Processing*, **8**(12), December 1999.

16. K. Lengwehasatit, A. Ortega, A. Basso and A. Reibman, 'A novel computationally scalable algorithm for motion estimation', *Proc. VCIP98*, San Jose, January 1998.

17. V. G. Moshnyaga, 'A new computationally adaptive formulation of block-matching motion estimation', *IEEE Trans. CSVT*, **11**(1), January 2001.

18. V. K. Goyal and M. Vetterli, 'Computation-distortion characteristics of block transform coding', *Proc. ICASSP97*, Munich, April 1997.

# 11

# Transmission of Coded Video

## 11.1  INTRODUCTION

A video communication system transmits coded video data across a channel or network and the transmission environment has a number of implications for encoding and decoding of video. The capabilities and constraints of the channel or network vary considerably, for example from low bit rate, error-prone transmission over a mobile network to high bit rate, reliable transmission over a cable television network. Transmission constraints should be taken into account when designing or specifying video CODECs; the aim is not simply to achieve the best possible compression but to develop a video coding system that is well matched to the transmission environment.

This problem of 'matching' the application to the network is often described as a 'quality of service' (QoS) problem. There are two sides to the problem: the QoS *required* by the application (which relates to visual quality perceived by the user) and the QoS *offered* by the transmission channel or network (which depends on the capabilities of the network). In this chapter we examine QoS from these two points of view and discuss design approaches that help to match the offered and required QoS. We describe two examples of transmission scenarios and discuss how these scenarios influence video CODEC design.

## 11.2  QUALITY OF SERVICE REQUIREMENTS AND CONSTRAINTS

### 11.2.1  QoS Requirements for Coded Video

Successful transmission of coded video places a number of demands on the transmission channel or network. The main requirements ('QoS requirements') for real-time video transmission are discussed below.

*Data rate*

A video encoder produces coded video at a variable or constant rate (as discussed in Chapter 10). The key parameters for transmission are the *mean* bit rate and the *variation* of the bit rate.

The mean rate (or the constant rate for CBR video) depends on the characteristics of the source video (frame size, number of bits per sample, frame rate, amount of motion, etc.) and on the compression algorithm. Practical video coding algorithms incorporate a degree of compression control (e.g. quantiser step size and mode selection) that allows some control of the mean rate after encoding. However, for a given source (with a particular frame size and frame rate) there is an upper and lower limit on the achievable mean compressed bit rate. For example, 'broadcast TV quality' video (approximately $704 \times 576$ pixels per frame, 25 or 30 frames per second) encoded using MPEG-2 requires a mean encoded bit rate of around 2–5 Mbps for acceptable visual quality. In order to successfully transmit video at 'broadcast' quality, the network or channel must support at least this bit rate.

Chapter 10 explained how the variation in coded bit rate depends on the video scene content and on the type of rate control algorithm used. Without rate control, a video CODEC tends to generate more encoded data when the scene contains a lot of spatial detail and movement and less data when the scene is relatively static. Different encoding modes (such as I, P or B-pictures in MPEG video) produce varying amounts of coded data. An output buffer together with a rate control algorithm may be used to 'map' this variable rate to either a constant bit rate (no bit rate variation) or a variable bit rate with constraints on the maximum amount of variation.

## Distortion

Most of the practical algorithms for encoding of real-time video are lossy, i.e. some distortion is introduced by encoding and the decoded video sequence is not identical to the original video sequence. The amount of distortion that is acceptable depends on the application.

### Example 1

A movie is displayed on a large, high-quality screen at HDTV resolution. Capture and editing of the video material is of a very high quality and the viewing conditions are good. In this example, there is likely to be a low 'threshold' for distortion introduced by the video CODEC, since any distortion will tend to be highlighted by the quality of the material and the viewing conditions.

### Example 2

A small video 'window' is displayed on a PC as part of a desktop video-conferencing application. The scene being displayed is poorly lit; the camera is cheap and placed at an inconvenient angle; the video is displayed at a low resolution alongside a number of other application windows. In this example, we might expect a relatively high threshold for distortion. Because of the many other factors limiting the quality of the visual image, distortion introduced by the video CODEC may not be obvious until it reaches significant levels.

Ideally, distortion due to coding should be negligible, i.e. the decoded video should be indistinguishable from the original (uncoded) video. More practical requirements for distortion may be summarised as follows:

1. Distortion should be 'acceptable' for the application. As discussed above, the definition of 'acceptable' varies depending on the transmission and viewing scenario: distortion due to coding should preferably not be the dominant limiting factor for video quality.

2. Distortion should be near constant from a subjective viewpoint. The viewer will quickly become 'used' to a particular level of video quality. For example, analogue VHS video is relatively low quality but this has not limited the popularity of the medium because viewers accept a predictable level of distortion. However, sudden changes in quality (for example, 'blocking' effects due to rapid motion or distortion due to transmission errors) are obvious to the viewer and can have a very negative effect on perceived quality.

*Delay*

By its nature, real-time video is sensitive to delay. The QoS requirements in terms of delay depend on whether video is transmitted one way (e.g. broadcast video, streaming video, playback from a storage device) or two ways (e.g. video conferencing).

*Simplex* (one-way) video transmission requires frames of video to be presented to the viewer at the correct points in time. Usually, this means a constant frame rate; in the case where a frame is not available at the decoder (for example, due to frame skipping at the encoder), the other frames should be delayed as appropriate so that the original temporal relationships between frames are preserved. Figure 11.1 shows an example: frame 3 from the original sequence is skipped by the encoder (because of rate constraints) and the frames arrive at the decoder in order 1, 2, 4, 5, 6. The decoder must 'hold' frame 2 for two frame periods so that the later frames (4, 5, 6) are not displayed too early with respect to frames 1 and 2. In effect, the CODEC maintains a constant delay between capture and display of frames. Any accompanying media that is 'linked' to the video frames must remain synchronised: the most common example is accompanying audio, where a loss of synchronisation of more than about 0.1 s can be obvious to the viewer.

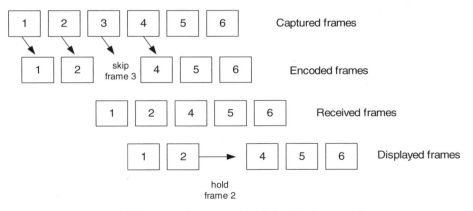

**Figure 11.1** Preserving temporal relationship between frames

*Duplex* (two-way) video transmission has the above requirements (constant delay in each direction, synchronisation between related media) plus the requirement that the total delay from capture to display must be kept low. A 'rule of thumb' for video conferencing is to keep the total delay less than 0.4 s. If the delay is longer than this, normal conversation becomes difficult and artificial.

Interactive applications, in which the viewer's actions affect the encoded video material, also have a requirement of low delay. An example is remote 'VCR' controls (play, stop, fast forward, etc.) for a streaming video source. A long delay between the user action (e.g. fast forward button) and the corresponding effect on the video source may make the application feel 'unresponsive'.

Figure 11.2 illustrates these three application scenarios.

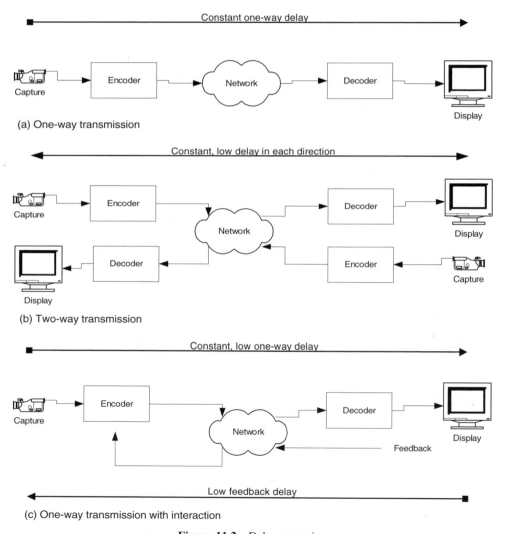

**Figure 11.2**   Delay scenarios

## 11.2.2 Practical QoS Performance

The previous section discussed the QoS *requirements* for coded video transmission; the other side of the equation is the QoS that can be *provided* by practical transmission environments.

*Data rate*

*Circuit-switched networks* such as the PSTN/POTS provide a constant bit rate connection. Examples include 33.6 kbps for a two-way modem connection over an analogue PSTN line; 56 kbps 'downstream' connection from an Internet service provider (ISP) over an analogue PSTN line; 128 kbps over basic rate ISDN.

*Packet-switched networks* such as Internet Protocol (IP) and Asynchronous Transfer Mode (ATM) provide a variable rate packet transmission service. This implies that these networks may be better suited to carrying coded video (with its inherently variable bit rate). However, the mean and peak packet transmission rate depend on the capacity of the network and may vary depending on the amount of other traffic in the network.

The data rate of a *digital subscriber line* connection (e.g. Asymmetric Digital Subscriber Line, ADSL) can vary depending on the quality of the line from the subscriber to the local PSTN exchange (the 'local loop'). The end-to-end rate achieved over this type of connection may depend on the 'core' network (typically IP) rather than the local ADSL connection speed.

*Dedicated transmission services* such as satellite broadcast, terrestrial broadcast and cable TV provide a constant bit rate connection that is matched to the QoS requirements of encoded television channels.

*Errors*

The circuit-switched PSTN and dedicated broadcast channels have a low rate of bit errors (randomly distributed, independent errors, case (a) in Figure 11.3). Packet-switched

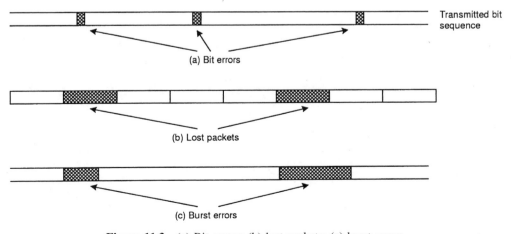

**Figure 11.3**  (a) Bit errors; (b) lost packets; (c) burst errors

networks such as IP usually have a low bit error rate but can suffer from packet loss during periods of network connection (loss of the data 'payload' of a complete network packet, case (b) in Figure 11.3). Packet loss is often 'bursty', i.e. a high rate of packet loss may be experienced during a particular period followed by a much lower rate of loss. Wireless networks (such as wireless LANs and personal communications networks) may experience high bit error rates due to poor propagation conditions. Fading of the transmitted signal can lead to bursts of bit errors in this type of network (case (c) in Figure 11.3, a sequence of bits containing a significant number of bit errors). Figure 11.4 shows the path loss (i.e. the variation in received signal power) between a base station and receiver in a mobile network, plotted as a function of distance. A mobile receiver can experience rapid fluctuations in signal strength (and hence in error rates) due to fading effects (such as the variation with distance shown in the figure).

*Delay*

Circuit-switched networks and dedicated broadcast channels provide a near-constant, predictable delay. Delay through a point-to-point wireless connection is usually predictable. The delay through a packet-switched network may be highly variable, depending on the route taken by the packet and the amount of other traffic. The delay through a network node, for example, increases if the traffic arrival rate is higher than the processing rate of the node. Figure 11.5 shows how two packets may experience very different delays as they traverse a packet-switched network. In this example, a packet following route A passes through four routers and experiences long queuing delays whereas a packet following route B passes through two routers with very little queuing time. (Some improvement may be gained by adopting virtual circuit switching where successive packets from the same source follow identical routes.) Finally, automatic repeat request (ARQ)-based error control can lead to very variable delays whilst waiting for packet retransmission, and so ARQ is not generally appropriate for real-time video transmission (except in certain special cases for error handling, described later).

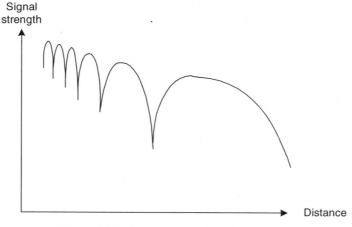

**Figure 11.4**  Path loss variation with distance

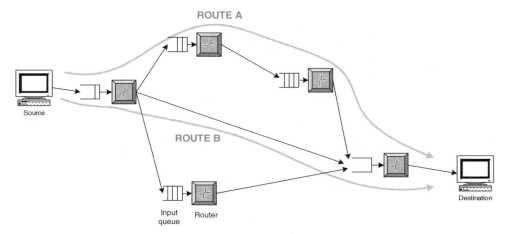

**Figure 11.5**   Varying delays within a packet-switched network

## 11.2.3   Effect of QoS Constraints on Coded Video

The practical QoS constraints described above can have a significant effect on the quality and performance of video applications.

*Data rate*

Most transmission scenarios require some form of rate control to adapt the inherently variable rate produced by a video encoder to a fixed or constrained bit rate supported by a network or channel. A rate control mechanism generally consists of an output buffer and a feedback control algorithm; practical rate control algorithms are described in Chapter 10.

*Errors*

A bit error in a compressed video sequence can cause a 'cascade' of effects that may lead to severe picture degradation. The following example illustrates the potential effects of a single bit error:

1. A single bit error occurs within variable-length coded transform coefficient data.

2. The coefficient corresponding to the affected VLC is incorrectly decoded. Depending on the magnitude of the coefficient, this may or may not have a visible effect on the decoded image. The incorrectly decoded coefficient may cause the current $8 \times 8$ block to appear distorted. If the current block is a luminance block, this will affect $8 \times 8$ pixels in the displayed image; if the current block contains chrominance data, this will affect $16 \times 16$ pixels (assuming $4:2:0$ sampling of the chrominance).

3. Subsequent VLCs may be incorrectly decoded because the error changes a valid VLC into another valid (but incorrect) VLC. In the worst case, the decoder may be unable to

Position of
original error

Propagation
to end of row

**Figure 11.6** Example of spatial error propagation

regain synchronisation with the correct sequence of syntax elements. The decoder can always recover at the next resynchronisation marker (such as a slice start code [MPEG-2] or GOB header [MPEG-4/H.263]). However, a whole section of the current frame may be corrupted before resynchronisation occurs. This effect is known as *spatial error propagation*, where a single error can cause a large spatial area of the frame to be distorted. Figure 11.6 shows an example: a single bit error affects a macroblock in the second-last row in this picture (coded using MPEG-4). Subsequent macroblocks are incorrectly decoded and the errored region propagates until the end of the row of macroblocks (where a GOB header enables the decoder to resynchronise).

4. If the current frame is used as a prediction reference (e.g. an I- or P-picture in MPEG or H.263), subsequent decoded frames are predicted from the distorted region. Thus an error-free decoded frame may be distorted due to an error in a previous frame (in decoding order): the error-free frame is decoded to produce a residual or difference frame which is then added to a distorted reference frame to produce a new distorted frame. This effect is *temporal error propagation* and is illustrated in Figure 11.7. The two frames

(a)                                                                   (b)

**Figure 11.7** Example of temporal error propagation

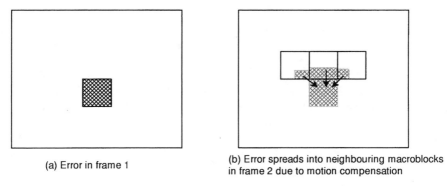

(a) Error in frame 1

(b) Error spreads into neighbouring macroblocks
in frame 2 due to motion compensation

**Figure 11.8**    Increase in errored area during temporal propagation

here were predicted from the errored frame shown in Figure 11.6: no further errors have occurred, but the distorted area continues to appear in further predicted frames. Because the macroblocks of the frames in Figure 11.7 are predicted using motion compensation, the errored region changes shape. The corrupted area may actually increase in subsequent predicted frames, as illustrated in Figure 11.8: in this example, motion vectors for macroblocks in frame 2 point 'towards' an errored area in frame 1 and so the error spreads out in frame 2. Over a long sequence of predicted frames, an errored region will tend to spread out and also to fade as it is 'added to' by successive correctly decoded residual frames.

In practice, packet losses are more likely to occur than bit errors in many situations. For example, a network transport protocol may discard packets containing bit errors. When a packet is lost, an entire section of coded data is discarded. A large section of at least one frame will be lost and this area may be increased due to spatial and temporal propagation.

*Delay*

Any delay within the video encoder and decoder must not cause the total delay to exceed the limits imposed by the application (e.g. a total delay of 0.4 s for video conferencing). Figure 11.9 shows the main sources of delay within a video coding application: each of the components shown (from the capture buffer through to the display buffer) introduces a delay. (Note that any multiplexing/packetising delay is assumed to be included in the encoder output buffer and decoder input buffer.)

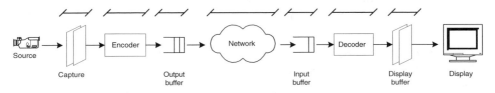

**Figure 11.9**    Sources of delay in a video CODEC application

The delay requirements place a constraint on certain aspects of the CODEC, including bidirectional prediction and output buffering. B-pictures (supported by the MPEG and H.263 standards) are predicted from two reference frames, one past, one future. The use of B-pictures introduces an extra delay of one frame in the decoder and at least one frame in the encoder (depending on the number of B-pictures between successive reference pictures) and so the improved compression efficiency of B-pictures needs to be balanced with delay constraints. A large encoder output buffer makes it easier to 'smooth' variations in encoded bit rate without rapid changes in quantisation (and hence quality): however, delay through the CODEC increases as the buffer size increases and so delay requirements limit the size of buffer that may be used.

## 11.3   DESIGN FOR OPTIMUM QoS

The problem of providing acceptable quality video over a channel with QoS constraints can be addressed by considering these constraints in the design and control of a video CODEC. There are a number of mechanisms within the video coding standards that may be exploited to maintain acceptable visual quality.

### 11.3.1   Bit Rate

Many different approaches to video rate control have been proposed and video quality may be maximised by careful choice of a rate control algorithm to match the type of video material and the channel characteristics. Chapter 10 discusses rate control in detail: the aims of a rate control algorithm (often conflicting) are to maximise quality within the bit rate and delay constraints of the channel and the application. Tools that may be used to achieve these aims include quantisation control, encoding mode selection and (if necessary) frame skipping. Further flexibility in the control of encoder bit rate is provided by some of the optional modes of H.263+ and MPEG-4, for example:

- Reference picture resampling (H.263+ Annex P) enables an encoder to change frame resolution 'on the fly'. With this optional mode, a picture encoded at the new resolution may be predicted from a resampled reference picture at the old resolution. Changing the spatial resolution can significantly change the encoded bit rate without interrupting the flow of frames.

- Object-based coding (MPEG-4) enables individual 'objects' within a video scene (for example, foreground and background) to be encoded largely independently. This can support flexible rate control by, for example, reducing the quality and frame update rate of less important background objects whilst maintaining high quality and update rate for visually significant foreground objects.

### 11.3.2   Error Resilience

Performance in the presence of errors can be improved at a number of stages in the CODEC 'chain',[1-3] including the following.

## Encoder

Resynchronisation methods may be used to limit error propagation. These include restart markers (e.g. slice start code, GOB header) to limit spatial error propagation, intra-coded pictures (e.g. MPEG I-pictures) to limit temporal error propagation and intra-coded macro-blocks to 'force' an error-free update of a region of the picture.[4] Splitting an encoded frame into sections that may be decoded independently limits the potential for error propagation and H.263+ Annex R (independent segment decoding) and the video packet mode of MPEG-4 support this. A further enhancement of error resilience is provided by the optional reversible variable length codes (RVLCs) supported by MPEG-4 and described in Chapter 8. Layered or scalable coding (such as the four scalable modes of MPEG-2) can improve performance in the presence of errors. The 'base' layer in a scalable coding algorithm is usually more sensitive to errors than the enhancement layer (s), and some improvement in error resilience has been demonstrated using *unequal error protection*, i.e. applying increased error protection to the base layer.[3]

## Channel

Suitable techniques include the use of error control coding[5,6] and 'intelligent' mapping of coded data into packets. The error control code specified in H.261 and H.263 (a BCH code) cannot correct many errors. More robust coding may be more appropriate (for example, concatenated Reed-Solomon and convolutional coding for MPEG-2 terrestrial or satellite transmission, see Section 11.4.1). Increased protection from errors can be provided at the expense of higher error correction overhead in transmitted packets. Careful mapping of encoded data into network packets can minimise the impact of a lost packet. For example, placing an independently decodeable unit (such as an independent segment or video packet) into each transmitted packet means that a lost packet will affect the smallest possible area of a decoded frame (i.e. the error will not propagate spatially beyond the data contained within the packet).

## Decoder

A transmission error may cause a 'violation' of the coded data syntax that is expected at the decoder. This violation indicates the approximate location of the corresponding errored region in the decoded frame. Once this is known, the decoder may implement *error concealment* to minimise the visual impact of the error. The extent of the errored region can be estimated once the position of the error is known, as the error will usually propagate spatially up to the next resynchronisation point (e.g. GOB header or slice start code). The decoder attempts to conceal the errored region by making use of spatially and temporally adjacent error-free regions. A number of error concealment algorithms exist and these usually take advantage of the fact that a human viewer is more sensitive to low-frequency components in the decoded image. An error concealment algorithm attempts to restore the low-frequency information and (in some cases) selected high-frequency information.

*Spatial error concealment* repairs the damaged region by interpolation from neighbouring error-free pixels.[1] Errors typically affect a 'stripe' of macroblocks across a picture (see for example Figure 11.6) and so the best method of interpolation is to use pixels immediately

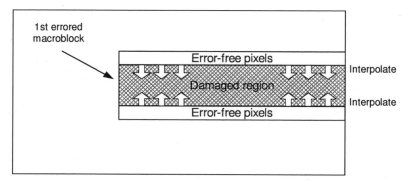

**Figure 11.10**   Spatial error concealment

above and below the damaged area as shown in Figure 11.10. A spatial filter may be used to 'smooth' the boundaries of the repaired area. More advanced concealment algorithms attempt to maintain significant features such as edges across the damaged region. This usually requires a computationally complex algorithm, for example using projection onto convex sets (see Section 9.3, 'Post-filtering').

*Temporal error concealment* copies data from temporally adjacent error-free frames to hide the damaged area.[7,8] A simple approach is to copy the same region from the previous frame (often available in the frame store memory at the decoder). A problem occurs when there is a change between the frames due to motion: the copied area appears to be 'offset' and this can be visually disturbing. This effect can be reduced by compensating for motion and this is straightforward if motion vectors are available for the damaged macroblocks. However, in many cases the motion vectors may be damaged themselves and must be reconstructed, for example by interpolating from the motion vectors of undamaged macro-blocks. Good results may be obtained by re-estimating the motion vectors in the decoder, but this adds significantly to the computational complexity.

Figure 11.11 shows how the error-resilient techniques described above may be applied to the encoder, channel and decoder.

## Combined approach

Recently, some promising methods for error handling involving cooperation between several stages of the transmission 'chain' have been proposed.[9-11] In a real-time video

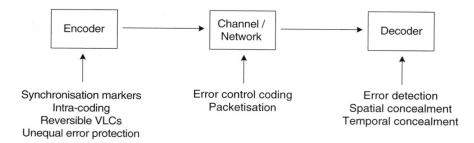

**Figure 11.11**   Application of error-resilient techniques

communication system it is not usually possible to retransmit damaged or lost packets due to delay constraints: however, it is possible for the decoder to signal the location of a lost packet to the encoder. The encoder can then determine the area of the frame that is likely to be affected by the error (a larger area than the original errored region due to motion-compensated prediction from the errored region) and encode macroblocks in this area using intra-coding. This will have the effect of 'cleaning up' the errored region once the feedback message is received. Alternatively, the technique of reference picture selection enables the encoder (and decoder) to choose an older, error-free frame for prediction of the next inter-frame once the position of the error is known. This requires both encoder and decoder to store multiple reference frames. The reference picture selection modes of H.263+ (Annex N and Annex U) may be used for this purpose.

These two methods of incorporating feedback are illustrated in Figures 11.12 and 11.13. In Figure 11.12, an error occurs during transmission of frame 1. The decoder signals the estimated location of the error to the encoder: meanwhile, the error propagates to frames 2 and 3 and spreads out due to motion compensation. The encoder estimates the likely spread of the damaged area and intra-codes an appropriate region of frame 4. The intra-coded macroblocks halt the temporal error propagation and 'clean up' decoded frames 4 and onwards. In Figure 11.13, an error occurs in frame 1 and the error is signalled back to the encoder by the decoder. On receiving the notification, the encoder selects a known 'good' reference frame (frame 0 in this case) to predict the next frame (frame 4). Frame 4 is inter-coded by motion-compensated prediction from frame 0 at the encoder. The decoder also selects frame 0 for reconstructing frame 4 and the result is an error-free frame 4.

### 11.3.3   Delay

The components shown in Figure 11.9 can each add to the delay (latency) through the video communication system:

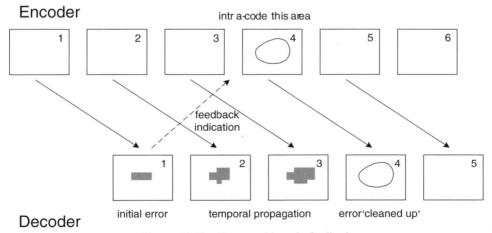

**Figure 11.12**   Error tracking via feedback

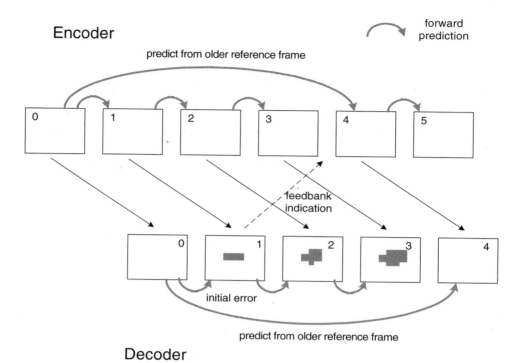

**Figure 11.13**    Reference picture selection

- *Capture buffer*: this should only add delay if the encoder 'stalls', i.e. it takes too long to encode incoming frames. This may occur in a software video encoder when insufficient processing capacity is available.

- *Encoder*: I- and P-frames do not introduce a significant delay: however, B-picture coding requires a multiple frame delay (as discussed in Chapter 4) and so the use of B-pictures should be limited in a delay-sensitive application.

- *Output buffer*: the output buffer adds a delay that depends on its maximum size (in bits). For example, if the channel bit rate is 64 kbps, a buffer of 32 kbits adds a delay of 0.5 s. Keeping the buffer small minimises delay, but makes it difficult to maintain consistent visual quality (as discussed in Chapter 10).

- *Network/channel*: if a resource reservation mechanism (such as those provided by RSVP [resource reservation protocol] in the Internet, see Section 11.4.2) is available, it may be possible to reserve a path with a guaranteed maximum delay. However, many practical networks cannot guarantee a particular delay through the network. The best alternative may be to use a 'low overhead' transport protocol such as the user datagram protocol (UDP), perhaps in conjunction with a streaming protocol such as the real time protocol (RTP) (see Section 11.4.2).

- *Input buffer*: the decoder input buffer size should be set to match the encoder output buffer. If the encoder and decoder are processing video at the same rate (i.e. the same number of frames per second), the decoder input buffer does not add any additional delay. (It can be shown[12] that the sum of the encoder buffer contents and decoder buffer contents is a constant if network delay is constant).

- *Decoder*: the use of B-pictures adds at most one frame's delay at the decoder and so this is not such a critical issue as at the encoder.

- *Display buffer*: as with the capture buffer, the display buffer should not add a significant delay unless a queue of decoded frames is allowed to build up due to variable decoding speed. In this case, the decoder should pause until the correct time for decoding a frame.

## 11.4 TRANSMISSION SCENARIOS

The design constraints and performance goals for a video CODEC are very dependent on the communications environment for which it is intended. Transmission scenarios for video communications applications range from high bit rate, high integrity transmission (e.g. television broadcasting) to low bit rate, unreliable environments (e.g. packet-based transmission over the Internet[13]). A number of 'framework' protocols have been developed to support video and audio transmission over different environments and some examples are listed in Table 11.1.

In this section we choose two popular transmission environments (digital television and LAN/Internet) and describe the protocols used for video transmission and their impact on the design of video CODECs.

### 11.4.1 Digital Television Broadcasting: MPEG-2 Systems/Transport

The MPEG-2 family of standards was developed with the aim of supporting 'television-quality' digital transmission of video and audio programmes. The video element is coded using MPEG-2 (Video) (described in Chapter 4) and the audio element is typically coded with MPEG-2 (Audio) Layer 3 ('MP3'). These elements are combined and transmitted via the MPEG-2 'Systems' framework.

MPEG-2 transmission is currently used in a number of environments including terrestrial radio transmission, direct broadcasting via satellite (DBS) and cable TV (CATV). MPEG-2 is also the chosen standard for video storage and playback on digital versatile disk (DVD). These transmission environments have a number of differences but they typically have some common characteristics: a fixed, 'guaranteed' bit rate, a predictable transmission delay and (usually) predictable levels of noise (and hence errors).

**Table 11.1** Transmission/storage environments and protocols

| Environment | Protocols | Notes |
|---|---|---|
| PSTN/ISDN | H.320,[14] H.324[15] | Constant bit rate, low delay networks |
| LAN/IP | H.323[16] | Variable packet rate, variable delay, unreliable transmission |
| Digital television broadcasting | MPEG-2 Systems[17] | Constant bit rate, error rates depend on transmission medium |

MPEG-2 (Systems) describes two methods of multiplexing audio, video and associated information, the program stream and the transport stream. In each case, streams of coded audio, video, data and system information are first packetised to form packetised elementary stream packets (*PES packets*).

## *The program stream*

This is the basic multiplexing method, designed for storage or distribution in a (relatively) error-free environment. A program stream carries a single *program* (e.g. a television programme) and consists of a stream of PES packets consisting of the video, audio and ancillary information needed to reconstruct the program. PES packets may be of variable length and these are grouped together in *packs*, each of which starts with a pack header.

Accurate timing control is essential for high-quality presentation of video and audio and this is achieved by a system of time references and time stamps. A decoder maintains a local system time clock (STC). Each pack header contains a system clock reference (SCR) field that is used to reset the decoder STC prior to decoding of the pack. PES packets contain time stamps and the decoder uses these to determine when the data in each packet should be decoded and presented. In this way, accurate synchronisation between the various data streams is achieved.

## *The transport stream*

The transport stream (TS) is designed for transmission environments that are prone to errors (such as terrestrial or satellite broadcast). The basic element of the TS is the PES packet. However, variable-length PES packets are further packetised to form fixed length TS packets (each is 188 bytes) making it easier to add error protection and identify and recover from transmission errors. A single TS may carry one or more programs multiplexed together. Figure 11.14 illustrates the way in which information is multiplexed into programs and then into TS packets.

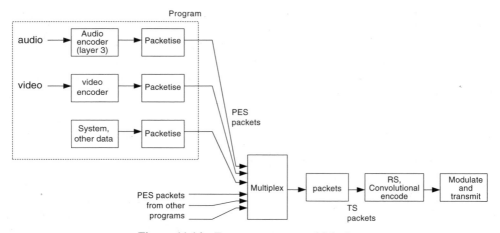

**Figure 11.14**  Transport stream multiplexing

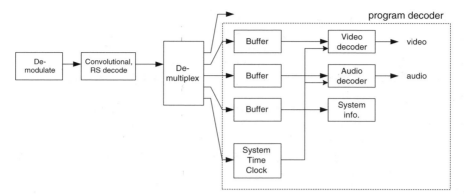

**Figure 11.15**   Transport stream demultiplexing

Two levels of error correcting coding provide protection from transmission errors. First, 16 parity bytes are added to each 188 byte TS packet to form a 204-byte Reed-Solomon codeword and the stream of codewords are further protected with a convolutional code (usually a 7/8 code, i.e. the encoder produces 8 output bits for every 7 input bits). The total error coding overhead is approximately 25%. The 'outer' convolutional code can correct random bit errors and the 'inner' Reed-Solomon code can correct burst errors up to 64 bits in length.

Figure 11.15 illustrates the process of demultiplexing and decoding an MPEG-2 TS. After correcting transmission errors, the stream of TS packets are demultiplexed and PES packets corresponding to a particular program are buffered and decoded. The decoder STC is periodically updated when a SCR field is received and the STC provides a timing reference for the video and audio decoders.

## *Implications for video CODEC design*

The characteristics of a typical MPEG-2 program are as follows:

- ITU-R 601 resolution video, 25 or 30 frames per second

- Stereo audio

- Video coded to approximately 3–5 Mbps

- Audio coded to approximately 300 kbps

- Total programme bit rate approximately 6 Mbps

An MPEG-2 video encoder/decoder design aims to provide high-quality video within these transmission parameters. The channel coding (Reed-Solomon and convolutional ECC) is designed to correct most transmission errors and error-resilient video coding is generally limited to simple error recovery (and perhaps concealment) at the decoder to handle the occasional uncorrected error. The STC and the use of time stamps in each PES packet provide accurate synchronisation.

## 11.4.2   Packet Video: H.323 Multimedia Conferencing

H.323 is an ITU-T 'umbrella' standard, describing a framework for multimedia communications over local area networks (LANs) and IP-based networks that do not support guaranteed QoS. Since its release in 1996, H.323 has gained popularity in Internet-based video and audio applications.

H.323 defines basic audio and video coding functionalities so that H.323-compliant devices and systems should be able to inter-operate with at least a minimum set of communication capabilities. H.323 provides independence from a particular network or platform (for example, by supporting translation between protocol frameworks for different network environments). It can assist with call set-up and management within a controlled 'zone' and it can support multi-point conferencing (three or more participants) and multi-cast (transmission from one source to many receivers).

### H.323 components

**Terminal**   This is the basic entity in an H.323-compliant system. An H.323 terminal consists of a set of protocols and functions and its architecture is shown in Figure 11.16. The mandatory requirements for an H.323 terminal (highlighted in the figure) are audio coding (using the G.711, G.723 or G.729 audio coding standards) and three control protocols: H.245 (channel control), Q.931 (call set-up and signalling) and registration/admission/status (RAS) (used to communicate with a gatekeeper, see below). Optional capabilities include video coding (H.261, H.263), data communications (using T.120) and the real time protocol (RTP) for packet transmission over IP networks. All H.323 terminals support point-to-point conferencing (i.e. one-to-one communications), support for multi-point conferencing (three or more participants) is optional.

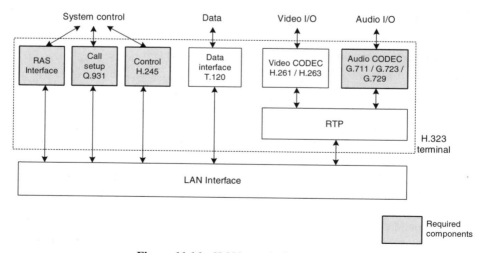

**Figure 11.16**   H.323 terminal architecture

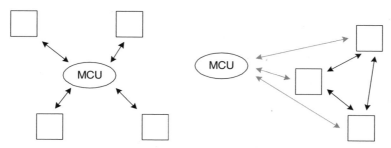

**Figure 11.17**  H.323 multi-point conferences

**Gateway**  An H.323 gateway provides an interface to other conferencing protocols such as H.320 (ISDN-based conferencing), H.324 (PSTN-based conferencing) and also analogue telephone handsets via the PSTN.

**Gatekeeper**  This is an optional H.323 entity that manages communications within a 'zone' (a defined set of H.323 components within the network). Gatekeeper functions include controlling the set-up and routeing of conferencing 'calls' within its zone. The gatekeeper can manage bandwidth usage within the zone by tracking the number of active calls and the bandwidth usage of each and barring new calls once the network has reached saturation.

**Multi-point control unit (MCU)**  This entity facilitates multi-point conferencing within H.323. The two main types of multi-point conference are *centralised* and *decentralised* (shown in Figure 11.17). In a centralised conference, all calls in the conference are routed through the MCU: hence each terminal only has to deal with point-to-point ('unicast') communications. This places a heavy processing burden on the MCU but is guaranteed to work with all H.323 terminals. A decentralised conference requires H.323 terminals that support 'multi-cast' communications: each terminal multi-casts its data to all other terminals in the conference and the MCU's role is to set up the conference and provide control and status information to each participant.

## Video coding in the H.323 environment

If an H.323 terminal supports video communication, it must be capable of using H.261 coding at QCIF resolution (see Chapter 5). Optionally, it may support H.263 coding and other resolutions (e.g. CIF, 4CIF). The capabilities of each terminal in a conference are signalled via the H.245 protocol: in a typical session, the terminals will choose the 'lowest common denominator' of video support. This could be H.261 (the minimum support), H.263 (baseline) or H.263 with optional modes.

H.323 is becoming popular for communications over the Internet. The Internet is inherently unreliable and this influences the choice of video coding tools and transmission protocols. The basic transport protocol is the unreliable datagram protocol (*UDP*): packets are transmitted without acknowledgement and are not guaranteed to reach their destination. This keeps delay to a minimum but packets may arrive out of order, late or not at all.

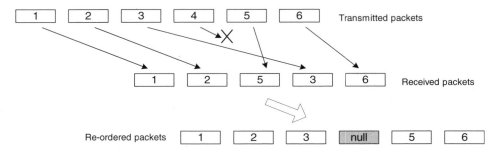

**Figure 11.18**  Packet sequencing using RTP

*RTP* may be used 'on top' of UDP for transmission of coded video and audio. RTP adds time stamps and sequence numbers to UDP packets, enabling a decoder to identify lost, delayed or out-of-sequence packets. If possible, a receiver will reorder the packets prior to decoding; if a packet does not arrive in time, its position is signalled to the decoder so that error recovery can be carried out. Figure 11.18 illustrates the way in which RTP reorders packets and signals the presence of lost packets. Packet 4 from the original sequence is lost during transmission and the remaining packets are received out of order. Sequence numbering and time stamps enable the packets to be reordered and indicate the absence of packet 4.

The *real time control protocol* (*RTCP*) may be used to monitor and control an RTP session. RTCP sends quality control messages to each participant in the session containing useful QoS information such as the packet loss rate.

The *resource reservation protocol* (*RSVP*) enables terminals to request a 'guaranteed' transmission bandwidth for the duration of the communication session. This improves the available QoS for real-time video and audio communications but requires support from every switch or router in the section of network traversed by the session.

### Implications for video CODEC design

Video coding for two-way conferencing in an H.323 environment should support low delay and low bit-rate coding. Coding tools such as B-pictures that add to encoding delay should probably be avoided. Depending on the packet loss rate (which may be signalled by the RTCP protocol), an encoder may choose to implement error-resilient features such as increased intra-coding and resynchronisation markers (to limit spatial and temporal error propagation) and the use of slice-structured coding (e.g. Annexes K and V of H.263) to map coded video to equal-sized packets. A video decoder can use the information contained within an RTP packet header to determine the exact presentation time of each decoded packet and to implement error handling and error concealment when a lost packet is detected.

## 11.5  SUMMARY

Successful video communications relies upon matching the QoS required by an application with the QoS provided by the transmission network. In this chapter we discussed key QoS parameters from the point of view of the video CODEC and the network. Removing subjective and statistical redundancy through the video compression process has the disadvantage that the compressed data becomes sensitive to transmission impairments such as delays and errors. An effective solution to the QoS problem is to deal with it both in the video CODEC (for example by introducing error-resilient features and matching the rate control algorithm to the channel) and in the network (for example by adopting protocols such as RTP). We described two popular transmission scenarios, digital television broadcast and IP video conferencing, and their influence on video CODEC design. The result of taking the transmission environment into account is a distinctly different CODEC in each case.

Video CODEC design is also heavily influenced by the implementation platform and in the next chapter we discuss alternative platforms and their implications for the designer.

# REFERENCES

1. Y. Wang, S. Wenger, J. Wen and A. Katsaggelos, 'Review of error resilient coding techniques for real-time video communications', *IEEE Signal Processing Magazine*, July 2000.
2. B. Girod and N. Färber, 'Error-resilient standard-compliant video coding', from *Recovery Techniques for Image and Video Compression*, Kluwer Academic Publishers, 1998.
3. I. E. G. Richardson, 'Video coding for reliable communications', Ph.D. thesis, Robert Gordon University, 1999.
4. J. Y. Liao and J. Villasensor, 'Adaptive intra block update for robust transmission of H.263', *IEEE Trans. CSVT*, February 2000.
5. W. Kumwilaisak, J. Kim and C. Jay Kuo, 'Video transmission over wireless fading channels with adaptive FEC', *Proc. PCS01*, Seoul, April 2001.
6. M. Bystrom and J. Modestino, 'Recent advances in joint source-channel coding of video', *Proc. URSI Symposium on Signals, Systems and Electronics*, Pisa, Italy, 1998.
7. S. Tsekeridou and I. Pitas, 'MPEG-2 error concealment based on block-matching principles', *IEEE Trans. Circuits and Systems for Video Technology*, June 2000.
8. J. Zhang, J. F. Arnold and M. Frater, 'A cell-loss concealment technique for MPEG-2 coded video', *IEEE Trans. CSVT*, June 2000.
9. B. Girod and N. Färber, 'Feedback-based error control for mobile video transmission', *Proc. IEEE* (special issue on video for mobile multimedia), 1999.
10. P-C. Chang and T-H. Lee, 'Precise and fast error tracking for error-resilient transmission of H.263 video', *IEEE Trans. Circuits and Systems for Video Technology*, June 2000.
11. N. Färber, B. Girod and J. Villasensor, 'Extensions of the ITU-T recommendation H.324 for error-resilient video transmission', *IEEE Communications Magazine*, June 1998.
12. Y. Yang, 'Rate control for video coding and transmission', Ph.D. thesis, Cornell University, 2000.
13. G. J. Conklin et al., 'Video coding for streaming media delivery on the internet', *IEEE Trans. CSVT*, **11**(3), March 2001.
14. ITU-T Recommendation H.320, 'Line transmission of non-telephone signals', 1992.
15. ITU-T Recommendation H.324, 'Terminal for low bitrate multimedia communication', 1995.
16. ITU-T Recommendation H.323, 'Packet based multimedia communications systems', 1997.
17. ISO/IEC 13818-1, 'Information technology: generic coding of moving pictures and associated audio information: Systems', 1995.

# 12

# Platforms

## 12.1 INTRODUCTION

In the early days of video coding technology, systems tended to fall into two categories, dedicated hardware designs for real-time video coding (e.g. H.261 videophones) or software designs for 'off-line' (not real-time) image or video coding (e.g. JPEG compression/decompression software). The continued increases in processor performance, memory density and storage capacity have led to a blurring of these distinctions and video coding applications are now implemented on a wide range of processing platforms. General-purpose platforms such as personal computer (PC) processors can achieve respectable real-time coding performance and benefit from economies of scale (i.e. widespread availability, good development tools, competitive cost). There is still a need for dedicated hardware architectures in certain niche applications, such as high-end video encoding or very low power systems. The 'middle ground' between the general-purpose platform and the dedicated hardware design (for applications that require more processing power than a general-purpose processor can provide but where a dedicated design is not feasible) was, until recently, occupied by programmable 'video processors'. So-called 'media processors', providing support for wider functionalities such as audio and communications processing, are beginning to occupy this market. There is currently a convergence of processing platforms, with media extensions and features being added to traditionally distinct processor families (embedded, DSP, general-purpose) so that the choice of platform for video CODEC designs is wider than ever before.

In this chapter we attempt to categorise the main platform alternatives and to compare their advantages and disadvantages for the designer of a video coding system. Of course, some of the information in this chapter will be out of date by the time this book is published, due to the rapid pace of development in processing platforms.

## 12.2 GENERAL-PURPOSE PROCESSORS

A desktop PC contains a processor that can be described as 'general-purpose'. The processor is designed to provide acceptable performance for a wide range of applications such as office, games and communications applications. Manufacturers need to balance the user's demand for higher performance against the need to keep costs down for a mass-market product. At the same time, the large economies of scale in the PC market make it possible for the major manufacturers to rapidly develop and release higher-performance versions of the processors. Table 12.1 lists some of the main players in the market and their recent processor offerings (as of August 2001).

**Table 12.1**   Popular PC processors

| Manufacturer | Latest offering | Features |
| --- | --- | --- |
| Intel | Pentium 4 | Clock speed up to 2 GHz; highly pipelined; 128-bit single instruction multiple data (SIMD) |
| Motorola | PowerPC G4 | Clock speed up to about 1 GHz; 128-bit vector processing |
| AMD | Athlon | Clock speed up to 1.4 GHz; multiple integer and floating-point execution units; SIMD processing |

## 12.2.1   Capabilities

PC processors can be loosely characterised as follows:

- good performance at running 'general' applications;
- not optimised for any particular class of application (though the recent trend is to add features such as SIMD capabilities to support multimedia applications);
- high power consumption (though lower-power versions of the above processors are available for mobile devices);
- support word lengths of 32 bits or more, fixed and floating point arithmetic;
- support for SIMD instructions (for example carrying out the same operation on $4 \times 32$-bit words).

The popular PC operating systems (Windows and Mac O/S) support multi-tasking applications and offer good support for external hardware (via plug-in cards or interfaces such as USB).

## 12.2.2   Multimedia Support

Recent trends towards multimedia applications have led to increasing support for real-time media. There are several 'frameworks' that may be used within the Windows O/S, for example, to assist in the rapid development and deployment of real-time applications. The DirectX and Windows Media frameworks provide standardised interfaces and tools to support efficient capture, processing, streaming and display of video and audio.

The increasing usage of multimedia has driven processor manufacturers to add architectural and instruction support for typical multimedia processing operations. The three processor families listed in Table 12.1 (Pentium, PowerPC, Athlon) each support a version of 'single instruction, multiple data' (SIMD) processing. Intel's MMX and SIMD extensions[1,2] provide a number of instructions aimed at media processing. A SIMD instruction operates on multiple data elements simultaneously (e.g. multiple 16-bit words within a 64-bit or 128-bit register). This facilitates computationally intensive, repetitive operations such

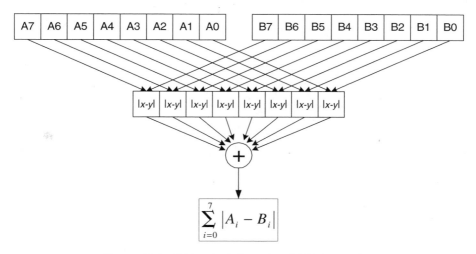

**Figure 12.1** SAD calculation using SIMD instruction

as motion estimation (e.g. calculating sum of absolute differences, SAD) and DCT (e.g. multiply–accumulate operations). Figure 12.1 shows how the Intel instruction `psadbw` may be used to calculate SAD for eight pairs of input samples ($A_i$, $B_i$) in parallel, leading to a potentially large computational saving.

Table 12.2 summarises the main advantages and disadvantages of PC platforms for video coding applications. The large user base and comprehensive development support make it an attractive platform for applications such as desktop video conferencing (Figure 12.2) in which a video CODEC is combined with a number of other components such as audio CODEC, chat and document sharing to provide a flexible, low-cost video communication system.

**Table 12.2** Advantages and disadvantages of PC platform

| Advantages | Disadvantages |
|---|---|
| High market penetration, very large potential user base | Computationally intensive video coding functions must be carried out in software |
| Availability of efficient compilers and powerful development tools | Medium to high power consumption |
| Multimedia extension functions to improve video processing performance | Use of 'special' instructions such as SIMD limits the portability of the video coding application |
| Efficient multi-tasking with other applications | Processor resources not always available (can be problematic for real-time video) |
| Availability of multimedia application development frameworks | |

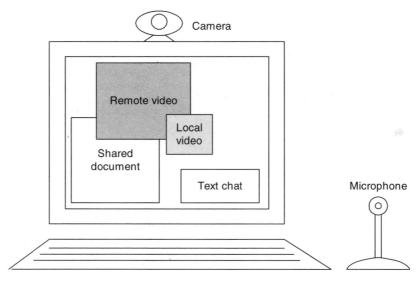

**Figure 12.2**   PC-based video conferencing

## 12.3   DIGITAL SIGNAL PROCESSORS

Digital signal processors (DSPs) are designed to efficiently handle applications that are based around computationally intensive signal processing algorithms. Typical applications include audio processing (e.g. filtering and compression), telecommunications functions (such as modem processing, filtering and echo cancellation) and signal conditioning (transformation, noise reduction, etc.). Mass-market applications for DSPs include PC modems, wireless and hand-held communications processing, speech coding and image processing. These applications typically require good signal processing performance in a power, cost and/or space-limited environment. DSPs can be characterised as follows:

- high performance for a selected range of signal processing operations;
- low/medium power consumption;
- low/medium cost;
- fixed or point arithmetic capability;
- limited on- and off-chip code and data storage (depending on the available address space);
- 16- or 32-bit wordlength.

Table 12.3 lists a few popular DSPs and compares their features: this is only a small selection of the wide range of DSPs on the market. As well as these discrete ICs, a number of manufacturers provide DSP cores (hardware architectures designed to be integrated with other modules on a single IC).

**Table 12.3** Popular DSPs

| Manufacturer | Device | Features |
| --- | --- | --- |
| Texas Instruments | C5000 series | Low power, 16-bit, up to 500 MIPS (instructions per second), optimised for portable devices and communications |
| | C6000 series | Medium power, 16 or 32-bit, 1000–4000 MIPS, fixed or floating-point arithmetic, optimised for broadband communications and image processing |
| Analog Devices | ADSP-218x and 219x series | Low power, 16-bit, over 300 MIPS |
| | SHARC | 32-bit, fixed and floating-point arithmetic, SIMD instructions, 600 MOPS (operations per second) |
| Motorola | DSP563xx | 24-bit, fixed-point, up to 200 MIPS, PCI bus interface |
| | DSP568xx | 16-bit, fixed-point, combines DSP and microcontroller features |

A key feature of DSPs is the ability to efficiently carry out repetitive processing algorithms such as filtering and transformation. This means that they are well suited to many of the computationally intensive functions required of a typical DCT-based video CODEC, such as motion estimation, DCT and quantisation, and some promising performance results have been reported.[3,4] Because a DSP is specifically designed for this type of application, this performance usually comes without the penalty of high power consumption. Support for related video processing functions (such as video capture, transmission and rendering) is likely to be limited. The choice of application development tools is not as wide as for the PC platform and high-level language support is often limited to the C language. Table 12.4 summarises the advantages and disadvantages of the DSP platform for video coding applications.

In a typical DSP development scenario, code is developed on a host PC in C, cross-compiled and downloaded to a development board for testing. The development board consists of a DSP IC together with peripherals such as memory, A/D converters and other interfaces. To summarise, a DSP platform can provide good performance with low power consumption but operating system and development support is often limited. DSPs may be a suitable platform for low-power, special-purpose applications (e.g. a hand-held videophone).

**Table 12.4** Advantages and disadvantages of DSP platform

| Advantages | Disadvantages |
| --- | --- |
| Low power consumption | Less well suited to 'higher-level' complex aspects of processing |
| Relatively high computational performance | Limited development support |
| Low price | Limited operating system support |
| Built-in telecommunications support (e.g. modem functions, A/D conversion) | Limited support for external devices (e.g. frame capture and display) |

## 12.4  EMBEDDED PROCESSORS

The term 'embedded processor' usually refers to a processor or controller that is "embedded" into a larger system, in order to provide programmable control and perhaps processing capabilities alongside more specialist, dedicated devices. Embedded processors are widely used in communications (mobile devices, network devices, etc) and control applications (e.g. automotive control). Typical characteristics are:

- low power consumption;
- low cost;
- limited processing and addressing capabilities;
- limited word lengths;
- fixed-point arithmetic.

Until recently, an embedded processor would not have been considered suitable for video coding applications because of severely limited processing capabilities. However, in common with other types of processor, the processing 'power' of new generations of embedded processor continues to increase. Table 12.5 summarises the features of some popular embedded processors.

The popular ARM and MIPS processors are licensed as cores for integration into third-party systems. ARM is actively targeting low-power video coding applications, demonstrating 15 frames per second H.263 encoding and decoding (QCIF resolution) on an ARM9[5] and developing co-processor hardware to further improve video coding performance.

Table 12.6 summarises the advantages and disadvantages of embedded processors for video coding applications. Embedded processors are of interest because of their large market penetration (for example, in the high-volume mobile telephone market). Running low-complexity video coding functions in software on an embedded processor (perhaps with limited dedicated hardware assistance) may be a cost-effective way of bringing video applications to mobile and wireless platforms. For example, the hand-held videophone is seen as a key application for the emerging '3G' high bit-rate mobile networks. Video coding on low-power embedded or DSP processors may be a key enabling technology for this type of device.

**Table 12.5**  Embedded processor features

| Manufacturer | Device | Features |
|---|---|---|
| MIPS | 4K series | Low power, 32-bit, up to approx. 400 MIPS, multiply–accumulate support (4KM) |
| ARM | ARM9 series | Low power, 16-bit, up to 220 MIPS |
| ARM/Intel | StrongARM series | Low power, 32-bit, up to 270 MIPS |

**Table 12.6**   Advantages and disadvantages of embedded processors

| Advantages | Disadvantages |
| --- | --- |
| Low power consumption | Limited performance |
| Low price | Limited word lengths, arithmetic, address spaces |
| High market penetration | (As yet) few features to support video processing |
| Good development tool support | |
| Increasing performance | |

## 12.5   MEDIA PROCESSORS

DSPs have certain advantages over general-purpose processors for video coding applications; so-called 'media processors' go a step further by providing dedicated hardware functions that support video and audio compression and processing. The general concept of a media processor is a 'core' processor together with a number of dedicated co-processors that carry out application-specific functions.

The architecture of the Philips TriMedia platform is shown in Figure 12.3. The core of the TriMedia architecture is a very long instruction word (VLIW) processor. A VLIW processor can carry out operations on multiple data words (typically four 32-bit words in the case of the TriMedia) at the same time. This is a similar concept to the SIMD instructions described earlier (see for example Figure 12.1) and is useful for video and audio coding applications. Computationally intensive functions in a video CODEC such as motion estimation and DCT may be efficiently carried out using VLIW instructions.

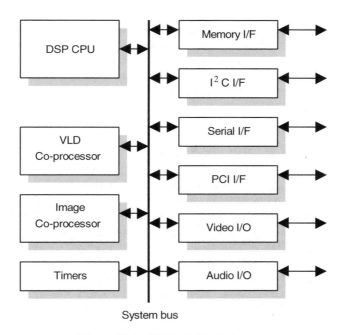

**Figure 12.3**   TriMedia block diagram

**Table 12.7**  Advantages and disadvantages of media processors

| Advantages | Disadvantages |
|---|---|
| Good performance for video coding | Application-specific features may not support future coding standards |
| Application-specific features (e.g. co-processors) | Good performance requires extensive code optimisation |
| High-level language support | Limited development tool support |
| Medium power consumption and cost | Limited market penetration |

The co-processors in the TriMedia architecture are designed to reduce the computational burden on the 'core' by carrying out intensive operations in hardware. Available co-processor units, shown in Figure 12.3, include video and audio interfaces, memory and external bus interfaces, an image co-processor and a variable-length decoder (VLD). The image co-processor is useful for pre- and post-processing operations such as scaling and filtering, and the VLD can decode an MPEG-2 stream in hardware (but does not currently support other coding standards). With careful software design and optimisation, a video coding application running on the TriMedia can offer good performance at a modest clock speed whilst retaining some of the benefits of a general-purpose processor (including the ability to program the core processor in C or C++ software).[6]

The MAP processor developed by Equator and Hitachi is another media processor that has generated interest recently. The heart of the processor is a VLIW core, surrounded by peripheral units that deal with video I/O, communications, video filtering and variable-length coding. According to the manufacturer, the MAP-CA can achieve impressive performance for video coding applications, for example encoding MPEG-2 Main Profile/Main Level video at 30 frames per second using 63% of the available processing resources.[7] This is higher than the reported performance of similar applications on the TriMedia.

Media processors have yet to capture a significant part of the market, and it is not yet clear whether the 'halfway house' between dedicated hardware and general-purpose software platforms will be a market winner. Table 12.7 summarises their main advantages and disadvantages for video coding.

## 12.6  VIDEO SIGNAL PROCESSORS

Video signal processors are positioned between media processors (which aim to process multiple media efficiently) and dedicated hardware CODECs (designed to deal with one video coding standard or a limited range of standards). A video signal processor contains dedicated units for carrying out common video coding functions (such as motion estimation, DCT/IDCT and VLE/VLD) but allows a certain degree of programmability, enabling a common platform to support a number of standards and to be at least partly 'future proof' (i.e. capable of supporting future extensions and new standards). An example is the VCPex offered by 8 × 8 Inc.: this is aimed at video coding applications (but also has audio coding support). The VCPex architecture (Figure 12.4) consists of two 32-bit data buses, labelled SRAM and DRAM. The SRAM bus is connected to the main controller (a RISC processor), a static RAM memory interface and other external interfaces. This 'side' of the VCPex deals

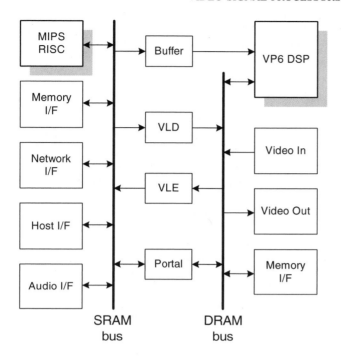

**Figure 12.4**   VCPex architecture

with lower bit-rate data such as compressed video, graphics and also coded audio. The
DRAM bus is connected to a dedicated video processor (the VP6), a dynamic RAM interface
and video input and output ports. The DRAM 'side' deals with high bit-rate, uncompressed
video and with most of the computationally intensive video coding operations. Variable-
length encoding and decoding are handled by dedicated VLE and VLD modules. This
partitioned architecture enables the VCPex to achieve good video coding and decoding
performance with relatively low power consumption. Computationally intensive video
coding functions (and pre- and post-processing) are handled by dedicated modules, but at
the same time the RISC and VP6 processors may be reprogrammed to support a range of
coding standards.

Table 12.8 summarises the advantages and disadvantages of this type of processor. Video
signal processors do not appear to be a strong force in the video communications market,

**Table 12.8**   Advantages and disadvantages of video signal processors

| Advantages | Disadvantages |
|---|---|
| Good performance for video coding | Application-specific features may not support future coding standards (but generally more flexible than dedicated hardware) |
| Application-specific features | Reprogramming likely to require high effort |
| Limited programmability | Limited development tool support |
| | Cost tends to be relatively high for mass market applications |
| | Dependent on a single manufacturer |

perhaps because they can be outperformed by a dedicated hardware design whilst they do not offer the same flexibility as a media processor or general-purpose processor.

## 12.7   CUSTOM HARDWARE

General-purpose processors and (to a lesser extent) media and video signal processors sacrifice a certain amount of performance in order to retain flexibility and programmability. A dedicated hardware design, optimised for a specific coding standard, is likely to offer the highest performance (in terms of video processing capacity and power consumption) at the expense of inflexibility.

The Zoran ZR36060 is a dedicated JPEG CODEC on a single chip capable of encoding or decoding ITU-R 601 video at 25 or 30 frames per second using Motion JPEG (see Chapter 4). A block diagram of the IC is shown in Figure 12.5. During encoding, video is captured by a dedicated video interface and stored in a 'strip buffer' that stores eight lines of samples prior to block processing. The JPEG core carries out JPEG encoding and the coded bit stream is passed to a first in first out (FIFO) buffer prior to output via the CODE interface. Decoding follows the reverse procedure. Control and status interfacing with a host processor is provided via the HOST interface. The chip is designed specifically for JPEG coding: however, some programmability of encoding and decoding parameters and quantisation tables is supported via the host interface.

Toshiba's TC35273 is a single-chip solution for MPEG-4 video and audio coding (Figure 12.6). Separate functional modules (on the left of the figure) handle MPEG-4 video coding and decoding (simple profile), audio coding and network communications, and each of these modules consists of a RISC controller and dedicated processing hardware. Video

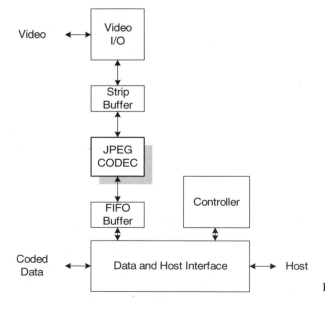

**Figure 12.5**   ZR36060 block diagram

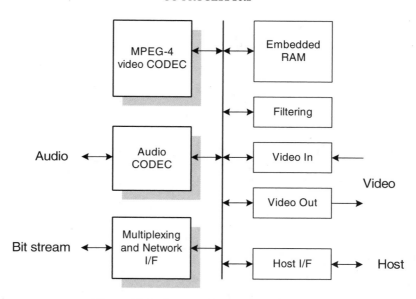

**Figure 12.6**  Toshiba TC35273 block diagram

capture, display and filtering are handled by co-processing modules. The IC is aimed at low-power, low bit-rate video applications and can handle QCIF video coding and decoding at 15 frames per second with a power consumption of 240 mW. A reduced-functionality version of this chip, the TC35274, handles only MPEG-4 video decoding.

Table 12.9 summarises the advantages and disadvantages of dedicated hardware designs. This type of CODEC is becoming widespread for mass market applications such as digital television receivers and DVD players. One potential disadvantage is the reliance on a single manufacturer in a specialist market; this is perhaps less likely to be a problem with general-purpose processors and media processors as they are targeted at a wider market.

## 12.8  CO-PROCESSORS

A co-processor is a separate unit that is designed to work with a host processor (such as a general-purpose PC processor). The co-processor (or 'accelerator') carries out certain computationally intensive functions in hardware, removing some of the burden from the host.

**Table 12.9**  Advantages and disadvantages of dedicated hardware CODECs

| Advantages | Disadvantages |
| --- | --- |
| High performance for video coding | No support for other coding standards |
| Optimised for target video coding standard | Limited control options |
| Cost-effective for mass-market applications | Dependent on a single manufacturer |

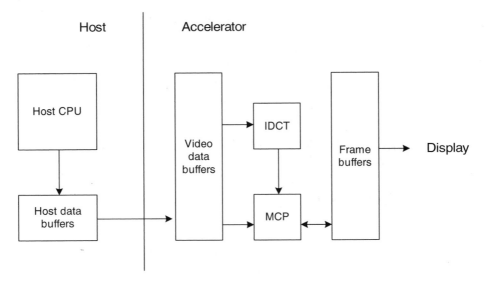

**Figure 12.7**   DirectX VA architecture

PC video display card manufacturers have begun to add support for common video coding functions to the display card hardware and a recent attempt to standardise the interface to this type of co-processor has led to the DirectX VA standard.[8] This aims to provide a standard API between a video decoding and display 'accelerator' and a host PC processor. The general architecture is shown in Figure 12.7. Complex, standard-specific functions such as variable-length decoding and header parsing are carried out by the host, whilst computationally intensive functions (but relatively regular and common to most standards) such as IDCT and motion compensation are 'offloaded' to the accelerator. The basic operation of this type of system is as follows:

1. The host decodes the bit stream and extracts rescaled block coefficients, motion vectors and header information.

2. This information is passed to the accelerator (using a standard API) via a set of data buffers.

3. The accelerator carries out IDCT and motion compensation and writes the reconstructed frame to a display buffer.

4. The display buffer is displayed on the PC screen and is also used as a prediction for further reconstructed frames.

Table 12.10 lists the advantages and disadvantages of this type of system. The flexibility of software programmability together with dedicated hardware support for key functions makes it an attractive option for PC-based video applications. Developers should benefit from the large PC market which will tend to ensure competitive pricing and performance for the technology.

**Table 12.10**  Advantages and disadvantages of co-processor architecture

| Advantages | Disadvantages |
|---|---|
| Flexible support for computationally intensive decoding functions | Dependent on specific platform and API |
| Supports all current DCT-based standards | Some intensive functions remain with host (e.g. VLD) |
| 'Front end' of decoder remains in software | Currently supports video decoding only |
| Large market for video display cards should lead to a number of alternative suppliers for this technology | |

## 12.9  SUMMARY

Table 12.11 attempts to compare the merits of the processing platforms discussed in this chapter. It should be emphasised that the rankings in this table are not exact and there will be exceptions in a number of cases (for example, a high-performance DSP that consumes more power than a media processor). However, the general trend is probably correct: the best coding performance per milliwatt of consumed power should be achievable with a dedicated hardware design, but on the other hand PC and embedded platforms are likely to offer the maximum flexibility and the best development support due to their widespread usage.

The recent trend is for a convergence between so-called 'dedicated' media processors and general-purpose processors, for example demonstrated by the development of SIMD/ VLIW-type functions for all the major PC processors. This trend is likely to continue as multimedia applications and services become increasingly important. At the same time, the latest generation of video coding standards (MPEG-4, H.263+ and H.26L) require relatively complex processing (e.g. to support object-based coding and coding mode decisions), as well as repetitive signal processing functions such as block-based motion estimation and

**Table 12.11**  Comparison of platforms (approximate)

| | Video coding performance | Power consumption | Flexibility | Development support |
|---|---|---|---|---|
| Best | Dedicated hardware | Dedicated hardware | PC processor | PC processor |
| | Video signal processor | Embedded processor | Embedded processor | Embedded processor |
| | Media processor | Digital signal processor | Digital signal processor | Digital signal processor |
| | PC processor | Video signal processor | Media processor | Media processor |
| | Digital signal processor | Media processor | Video signal processor | Video signal processor |
| Worst | Embedded processor | PC processor | Dedicated hardware | Dedicated hardware |

transform coding. These higher-level complexities are easier to handle in software than in dedicated hardware, and it may be that dedicated hardware CODECs will become less important (except for specialist, 'high-end' functions such as studio encoding) and that general-purpose processors will take care of mass-market video coding applications (perhaps with media processors or co-processors to handle low-level signal processing).

In the next chapter we examine the main issues that are faced by the designer of a software or hardware video CODEC, including issues common to both (such as interface requirements) and the separate design goals for a software or hardware CODEC.

# REFERENCES

1. M. Mittal, A. Peleg and U. Weiser, 'MMX technology architecture overview', *Intel Technology Journal*, 3rd Quarter, 1997.
2. J. Abel et al., 'Applications tuning for streaming SIMD extensions', *Intel Technology Journal*, 2nd Quarter, 1999.
3. H. Miyazawa, *H.263 Encoder: TMS320C6000 Implementation*, Texas Instruments Application Report SPRA721, December 2000.
4. K. Leung, N. Yung and P. Cheung, 'Parallelization methodology for video coding–an implementation on the TMS320C80', *IEEE Trans CSVT*, **10**(8), December 2000.
5. I. Thornton, *MPEG-4 over Wireless Networks*, ARM Inc. White Paper, 2000.
6. I. E. G. Richardson, K. Kipperman et al., 'Video coding using digital signal processors', *Proc. DSP World Conference*, Orlando, November 1999.
7. C. Basoglu et al., *The MAP-CA VLIW-based Media Processor*, Equator Technologies Inc. White Paper, January 2000. http://www.equator.com
8. G. Sullivan and C. Fogg, 'Microsoft Direct XVA: Video Acceleration API/DDI', Windows Platform Design Note, Microsoft, January 2000.

# 13

# Video CODEC Design

## 13.1  INTRODUCTION

In this chapter we bring together some of the concepts discussed earlier and examine the issues faced by designers of video CODECs and systems that interface to video CODECs. Key issues include interfacing (the format of the input and output data, controlling the operation of the CODEC), performance (frame rate, compression, quality), resource usage (computational resources, chip area) and design time. This last issue is important because of the fast pace of change in the market for multimedia communication systems. A short time-to-market is critical for video coding applications and we discuss methods of streamlining the design flow. We present design strategies for two types of video CODEC, a software implementation (suitable for a general-purpose processor) and a hardware implementation (for FPGA or ASIC).

## 13.2  VIDEO CODEC INTERFACE

Figure 13.1 shows the main interfaces to a video encoder and video decoder:

(a)  Encoder input: frames of uncompressed video (from a frame grabber or other source); control parameters.

(b)  Encoder output: compressed bit stream (adapted for the transmission network, see Chapter 11); status parameters.

(c)  Decoder input: compressed bit stream; control parameters.

(d)  Decoder output: frames of uncompressed video (send to a display unit); status parameters.

A video CODEC is typically controlled by a 'host' application or processor that deals with higher-level application and protocol issues.

### 13.2.1  Video In/Out

There are many options available for the format of uncompressed video into the encoder or out of the decoder and we list some examples here. (The four-character codes listed for options (a) and (b) are 'FOURCC' descriptors originally defined as part of the AVI video file format.)

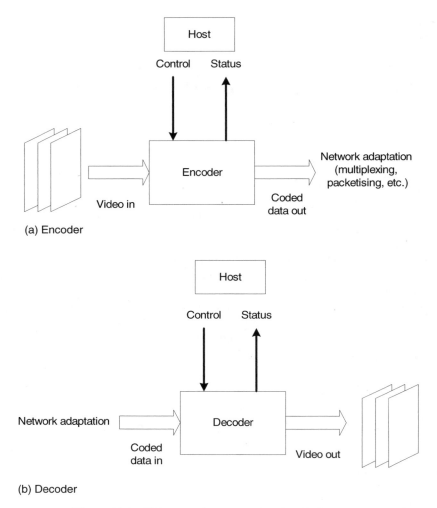

**Figure 13.1**   Video encoder (a) and decoder (b) interfaces

(a)   YUY2 (4 : 2 : 2). The structure of this format is shown in Figure 13.2. A sample of Y
(luminance) data is followed by a sample of Cb (blue colour difference), a second
sample of Y, a sample of Cr (red colour difference), and so on. The result is that the
chrominance components have the same vertical resolution as the luminance component but half the horizontal resolution (i.e. 4 : 2 : 2 sampling as described in Chapter 2).
In the example in the figure, the luminance resolution is $176 \times 144$ and the chrominance resolution is $88 \times 144$.

(b)   YV12 (4 : 2 : 0) (Figure 13.3). The luminance samples for the current frame are stored
in sequence, followed by the Cr samples and then the Cb samples. The Cr and Cb
samples have half the horizontal and vertical resolution of the Y samples. Each colour

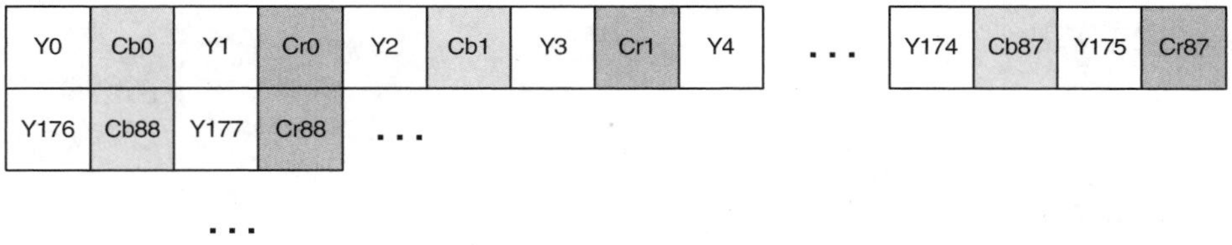

| Y0 | Cb0 | Y1 | Cr0 | Y2 | Cb1 | Y3 | Cr1 | Y4 | ... | Y174 | Cb87 | Y175 | Cr87 |

| Y176 | Cb88 | Y177 | Cr88 | ... |

...

**Figure 13.2**   YUY2 (4:2:2)

pixel in the original image maps to an average of 12 bits (effectively 1 Y sample, $\frac{1}{4}$ Cr sample and $\frac{1}{4}$ Cb sample), hence the name 'YV12'. Figure 13.4 shows an example of a frame stored in this format, with the luminance array first followed by the half-width and half-height Cr and Cb arrays.

(c)   Separate buffers for each component (Y, Cr, Cb). The CODEC is passed a pointer to the start of each buffer prior to encoding or decoding a frame.

As well as reading the source frames (encoder) and writing the decoded frames (decoder), both encoder and decoder require to store one or more reconstructed reference frames for motion-compensated prediction. These frame stores may be part of the CODEC (e.g. internally allocated arrays in a software CODEC) or separate from the CODEC (e.g. external RAM in a hardware CODEC).

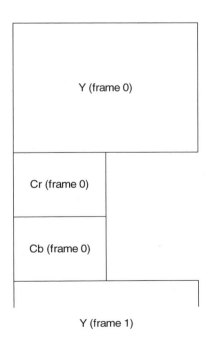

Y (frame 0)

Cr (frame 0)

Cb (frame 0)

Y (frame 1)

...

**Figure 13.3**   YV12 (4:2:0)

**Figure 13.4** Example of YV12 data

*Memory bandwidth* may be a particular issue for large frame sizes and high frame rates. For example, in order to encode or decode video at 'television' resolution (ITU-R 601, approximately $576 \times 704$ pixels per frame, 25 or 30 frames per second), the encoder or decoder video interface must be capable of transferring 216 Mbps. The data transfer rate may be higher if the encoder or decoder stores reconstructed frames in memory external to the CODEC. If forward prediction is used, the encoder must transfer data corresponding to three complete frames for each encoded frame, as shown in Figure 13.5: reading a new input frame, reading a stored frame for motion estimation and compensation and writing a reconstructed frame. This means that the memory bandwidth at the encoder input is at least $3 \times 216 = 648$ Mbps for ITU-R 601 video. If two or more prediction references are used for motion estimation/compensation (for example, during MPEG-2 B-picture encoding), the memory bandwidth is higher still.

## 13.2.2 Coded Data In/Out

Coded video data is a continuous sequence of bits describing the syntax elements of coded video, such as headers, transform coefficients and motion vectors. If modified Huffman coding is used, the bit sequence consists of a series of variable-length codes (VLCs) packed together; if arithmetic coding is used, the bits describe a series of fractional numbers each

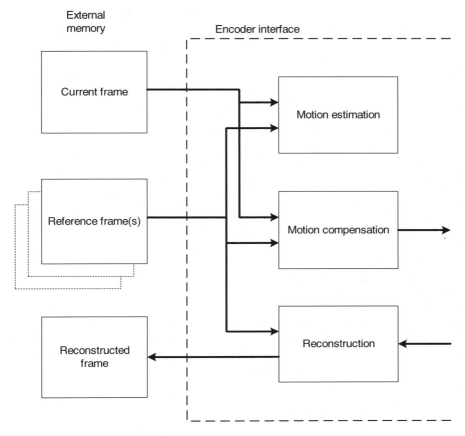

**Figure 13.5**  Memory access at encoder input

representing a series of data elements (see Chapter 8). The sequence of bits must be mapped to a suitable data unit for transmission/transport, for example:

1. *Bits*: If the transmission channel is capable of dealing with an arbitrary number of bits, no special mapping is required. This may be the case for a dedicated serial channel but is unlikely to be appropriate for most network transmission scenarios.

2. *Bytes or words*: The bit sequence is mapped to an integral number of bytes (8 bits) or words (16 bits, 32 bits, 64 bits, etc.). This is appropriate for many storage or transmission scenarios where data is stored in multiples of a byte. The end of the sequence may require to be padded in order to make up an integral number of bytes.

3. *Complete coded unit*: Partition the coded stream along boundaries that make up coded units within the video syntax. Examples of these coded units include slices (sections of a coded picture in MPEG-1, MPEG-2, MPEG-4 or H.263+), GOBs (groups of blocks, sections of a coded picture in H.261 or H.263) and complete coded pictures. The integrity of the coded unit is preserved during transmission, for example by placing each coded unit in a network packet.

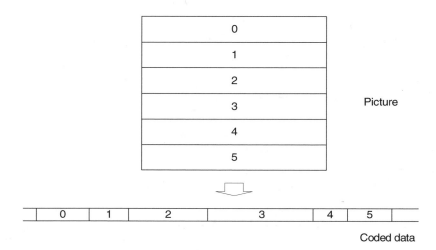

**Figure 13.6**   GOB locations in a frame and variable-size coded units

Figure 13.6 shows the locations of GOBs in a frame coded using H.263/MPEG-4. The coded units (GOBs in this case) correspond to regular areas of the original frame: however, when encoded, each GOB generates a different number of coded bits (due to variations in content within the frame). The result is that the GOBs generate the variable-size coded units shown in Figure 13.6.

An alternative is to use irregular-sized slices (e.g. using the slice structured mode in H.263+, video packet mode in MPEG-4). Figure 13.7 shows slice boundaries that cover irregular numbers of macroblocks in the original frame and are chosen such that, when coded, each slice contains a similar number of coded bits.

### 13.2.3   Control Parameters

Some of the more important control parameters are listed here (CODEC application programming interfaces [APIs] might not provide access to all of these parameters).

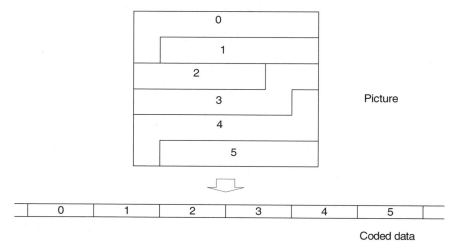

**Figure 13.7**   Slice boundaries in a picture and constant-size coded units

*Encoder*

**Frame rate** May be specified as a number of frames per second or as a proportion of frames to skip during encoding (e.g. skip every second frame). If the encoder is operating in a rate- or computation-constrained environment (see Chapter 10), then this will be a target frame rate (rather than an absolute rate) that may or may not be achievable.

**Frame size** For example, a 'standard' frame size (QCIF, CIF, ITU-R 601, etc) or a non-standard size.

**Target bit rate** Required for encoders operating in a rate-controlled environment.

**Quantiser step size** If rate control is not used, a fixed quantiser step size may be specified: this will give near-constant video quality.

**Mode control** For example 'inter' or 'intra' coding mode.

**Optional mode selection** MPEG-2, MPEG-4 and H.263 include a number of optional coding modes (for improved coding efficiency, improved error resilience, etc.). Most CODECs will only support a subset of these modes, and the choice of optional modes to use (if any) must be signalled or negotiated between the encoder and the decoder.

**Start/stop encoding** A series of video frames.

*Decoder*

Most of the parameters listed above are signalled to the decoder within the coded bit stream itself. For example, quantiser step size is signalled in frame/picture headers and (optionally) macroblock headers; frame rate is signalled by means of a timing reference in each picture header; mode selection is signalled in the picture header; and so on. Decoder control may be limited to 'start/stop'.

## 13.2.4 Status Parameters

There are many aspects of CODEC operation that may be useful as status parameters returned to the host application. These may include:

- actual frame rate (may differ from the target frame rate in rate- or computation-constrained environments);

- number of coded bits in each frame;

- macroblock mode statistics (e.g. number of intra/inter-macroblocks);

- quantiser step size for each macroblock (this may be useful for post-decoder filtering, see Chapter 9);

- distribution of coded bits (e.g. proportion of bits allocated to coefficients, motion vector data, header data);

- error indication (returned by the decoder when a transmission error has been detected, possibly with the estimated location of the error in the decoded frame).

## 13.3  DESIGN OF A SOFTWARE CODEC

In this section we describe the design goals and the main steps required to develop a video CODEC for a software platform.

### 13.3.1  Design Goals

A real-time software video CODEC has to operate under a number of constraints, perhaps the most important of which are computational (determined by the available processing resources) and bit rate (determined by the transmission or storage medium). Design goals for a software video CODEC may include:

1. Maximise encoded frame rate. A suitable target frame rate depends on the application, for example, 12–15 frames per second for desktop video conferencing and 25–30 frames per second for television-quality applications.

2. Maximise frame size (spatial dimensions).

3. Maximise 'peak' coded bit rate. This may seem an unusual goal since the aim of a CODEC is to compress video: however, it can be useful to take advantage of a high network transmission rate or storage transfer rate (if it is available) so that video can be coded at a high quality. Higher coded bit rates place higher demands on the processor.

4. Maximise video quality (for a given bit rate). Within the constraints of a video coding standard, there are usually many opportunities to 'trade off' video quality against computational complexity, such as the variable complexity algorithms described in Chapter 10.

5. Minimise delay (latency) through the CODEC. This is particularly important for two-way applications (such as video conferencing) where low delay is essential.

6. Minimise compiled code and/or data size. This is important for platforms with limited available memory (such as embedded platforms). Some features of the popular video coding standards (such as the use of B-pictures) provide high compression efficiency at the expense of increased storage requirement.

7. Provide a flexible API, perhaps within a standard framework such as DirectX (see Chapter 12).

8. Ensure that code is robust (i.e. it functions correctly for any video sequence, all allowable coding parameters and under transmission error conditions), maintainable and easily upgradeable (for example to add support for future coding modes and standards).

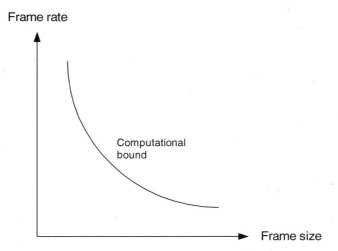

**Figure 13.8**  Trade-off of frame size and frame rate in a software CODEC

9. Provide platform independence where possible. 'Portable' software that may be executed on a number of platforms can have advantages for development, future migration to other platforms and marketability. However, achieving maximum performance may require some degree of platform-specific optimisation (such as the use of SIMD/VLIW instructions).

The first four design goals listed above may be mutually exclusive. Each of the goals (maximising frame rate, frame size, peak bit rate and video quality) requires an increased allocation of processing resources. A software video CODEC is usually constrained by the available processing resources and/or the available transmission bit rate. In a typical scenario, the number of macroblocks of video that a CODEC can process is roughly constant (determined by either the available bit rate or the available processing resources). This means that increased frame rate can only be achieved at the expense of a smaller frame size and vice versa. The graph in Figure 13.8 illustrates this trade-off between frame size and frame rate in a computation-constrained scenario. It may, however, be possible to 'shift' the line to the right (i.e. increase frame rate without reducing frame size or vice versa) by making better use of the available computational resources.

## 13.3.2  Specification and Partitioning

Based on the requirements of the syntax (for example, MPEG-2, MPEG-4 or H.263), an initial partition of the functions required to encode and decode a frame of video can be made. Figure 13.9 shows a simplified flow diagram for a block/macroblock-based inter-frame encoder (e.g. MPEG-1, MPEG-2, H.263 or MPEG-4) and Figure 13.10 shows the equivalent decoder flow diagram.

The order of some of the operations is fixed by the syntax of the coding standards. It is necessary to carry out DCT and quantisation of each block within a macroblock before generating the VLCs for the macroblock header: this is because the header typically contains a 'coded block pattern' field that indicates which of the six blocks actually contain coded transform coefficients. There is greater flexibility in deciding the order of some of the other

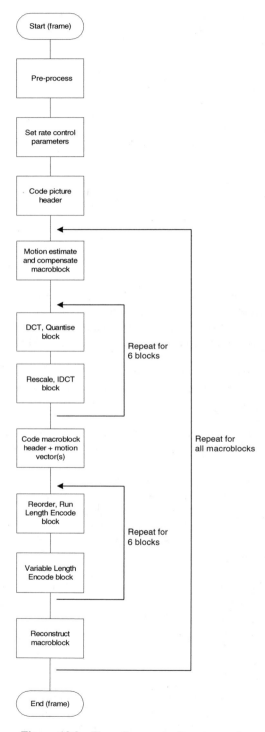

**Figure 13.9**   Flow diagram: software encoder

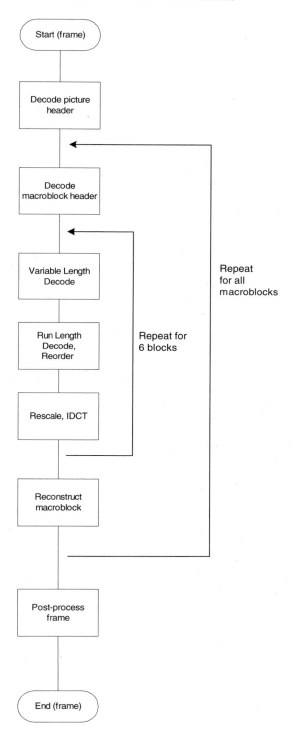

**Figure 13.10**  Flow diagram: software decoder

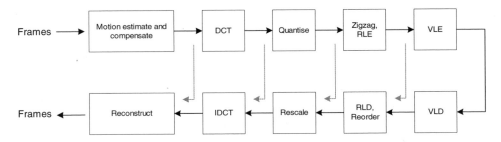

**Figure 13.11**   Encoder and decoder interoperating points

operations. An encoder may choose to carry out motion estimation and compensation for the entire frame before carrying out the block-level operations (DCT, quantise, etc.), instead of coding the blocks immediately after motion compensating the macroblock. Similarly, an encoder or decoder may choose to reconstruct each motion-compensated macroblock either immediately after decoding the residual blocks or after the entire residual frame has been decoded.

The following principles can help to decide the structure of the software program:

1. Minimise interdependencies between coding functions in order to keep the software modular.

2. Minimise data copying between functions (since each copy adds computation).

3. Minimise function-calling overheads. This may involve combining functions, leading to less modular code.

4. Minimise latency. Coding and transmitting each macroblock immediately after motion estimation and compensation can reduce latency. The coded data may be transmitted immediately, rather than waiting until the entire frame has been motion-compensated before coding and transmitting the residual data.

### 13.3.3   Designing the Functional Blocks

A good approach is to start with the simplest possible implementation of each algorithm (for example, the basic form of the DCT shown in Equation 7.1) in order to develop a functional CODEC as quickly as possible. The first 'pass' of the design will result in a working, but very inefficient, CODEC and the performance can then be improved by replacing the basic algorithms with 'fast' algorithms. The first version of the design may be used as a 'benchmark' to ensure that later, faster versions still meet the requirements of the coding standard.

Designing the encoder and decoder in tandem and taking advantage of 'natural' points at which the two designs can interwork may further ease the design process. Figure 13.11 shows some examples of interworking points. For example, the residual frame produced after encoder motion compensation may be 'fed' to the decoder motion reconstruction function and the decoder output frame should match the encoder input frame.

## 13.3.4  Improving Performance

Once a basic working CODEC has been developed, the aim is to improve the performance in order to meet the design goals discussed above. This may involve some or all of the following steps:

1. Carry out software profiling to measure the performance of individual functions. This is normally carried out automatically by the compiler inserting timing code into the software and measuring the amount of time spent within each function. This process identifies 'critical' functions, i.e. those that take the most execution time.

2. Replace critical functions with 'fast' algorithms. Typically, functions such as motion estimation, DCT and variable-length coding are computationally critical. The choice of 'fast' algorithm depends on the platform and to some extent the design structure of the CODEC. It is often good practice to compare several alternative algorithms and to choose the best.

3. Unroll loops. See Section 6.8.1 for an example of how a motion estimation function may be redesigned to reduce the overhead due to incrementing a loop counter.

4. Reduce data interdependencies. Many processors have the ability to execute multiple operations in parallel (e.g. using SIMD/VLIW instructions); however, this is only possible if the operations are working on independent data.

5. Consider combining functions to reduce function calling overheads and data copies. For example, a decoder carries out inverse zigzag ordering of a block followed by inverse quantisation. Each operation involves a movement of data from one array into another, together with the overhead of calling and returning from a function. By combining the two functions, data movement and function calling overhead is reduced.

6. For computationally critical operations (such as motion estimation), consider using platform-specific optimisations such as inline assembler code, compiler directives or platform-specific library functions (such as Intel's image processing library).

Applying some or all of these techniques can dramatically improve performance. However, these approaches can lead to increased design time, increased compiled code size (for example, due to unrolled loops) and complex software code that is difficult to maintain or modify.

*Example*

An H.263 CODEC was developed for the TriMedia TM1000 platform.[1] After the 'first pass' of the software design process (i.e. without detailed optimisation), the CODEC ran at the unacceptably low rate of 2 CIF frames per second. After reorganising the software (combining functions and removing interdependencies between data), execution speed was increased to 6 CIF frames per second. Applying platform-specific optimisation of critical functions (using the TriMedia VLIW instructions) gave a further increase to 15 CIF frames per second (an acceptable rate for video-conferencing applications).

### 13.3.5   Testing

In addition to the normal requirements for software testing, the following areas should be checked for a video CODEC design:

- Interworking between encoder and decoder (if both are being developed).

- Performance with a range of video material (including 'live' video if possible), since some 'bugs' may only show up under certain conditions (for example, an incorrectly decoded VLC may only occur occasionally).

- Interworking with third-party encoder(s) and decoder(s). Recent video coding standards have software 'test models' available that are developed alongside the standard and provide a useful reference for interoperability tests.

- Decoder performance under error conditions, such as random bit errors and packet losses.

To aid in debugging, it can be useful to provide a 'trace' mode in which each of the main coding functions records its data to a log file. Without this type of mode, it can be very difficult to identify the cause of a software error (say) by examining the stream of coded bits. A real-time test framework which enables 'live' video from a camera to be coded and decoded in real time using the CODEC under development can be very useful for testing purposes, as can be bit-stream analysis tools (such as 'MPEGTool') that provide statistics about a coded video sequence.

Some examples of efficient software video CODEC implementations have been discussed.[2-7] Opportunities have been examined for parallelising video coding algorithms for multiple-processor platforms,[5-7] and a method has been described for splitting a CODEC implementation between dedicated hardware and software.[8] In the next section we will discuss approaches to designing dedicated VLSI video CODECs.

## 13.4   DESIGN OF A HARDWARE CODEC

The design process for a dedicated hardware implementation is somewhat different, though many of the design goals are similar to those for a software CODEC.

### 13.4.1   Design Goals

Design goals for a hardware CODEC may include:

1. Maximise frame rate.

2. Maximise frame size.

3. Maximise peak coded bit rate.

4. Maximise video quality for a given coded bit rate.

5. Minimise latency.

6. Minimise gate count/design 'area', on-chip memory and/or power consumption.

7. Minimise off-chip data transfers ('memory bandwidth') as these can often act as a performance 'bottleneck' for a hardware design.

8. Provide a flexible interface to the host system (very often a processor running higher-level application software).

In a hardware design, trade-offs occur between the first four goals (maximise frame rate/frame size/peak bit rate/quality) and numbers (6) and (7) above (minimise gate count/power consumption and memory bandwidth). As discussed in Chapters 6–8, there are many alternative architectures for the key coding functions such as motion estimation, DCT and variable-length coding, but higher performance often requires an increased gate count. An important constraint is the *cycle budget* for each coded macroblock. This can be calculated based on the target frame rate and frame size and the clock speed of the chosen platform.

*Example*

Target frame size:     QCIF (99 macroblocks per frame, H.263/MPEG-4 coding)
Target frame rate:     30 frames per second
Clock speed:     20 MHz

Macroblocks per second:     $99 \times 30 = 2970$
Clock cycles per macroblock:     $20 \times 10^6 / 2970 = 6374$

This means that all macroblock operations must be completed within 6374 clock cycles. If the various operations (motion estimation, compensation, DCT, etc.) are carried out serially then the sum total for all operations must not exceed this figure; if the operations are pipelined (see below) then any one operation must not take more than 6374 cycles.

## 13.4.2 Specification and Partitioning

The same sequence of operations listed in Figures 13.9 and 13.10 need to be carried out by a hardware CODEC. Figure 13.12 shows an example of a decoder that uses a 'common bus'

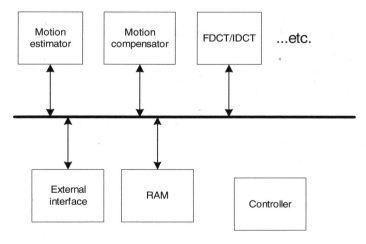

**Figure 13.12** Common bus architecture

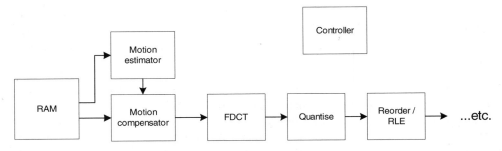

**Figure 13.13**  Pipelined architecture

architecture. This type of architecture may be flexible and adaptable but the performance
may be constrained by data transfer over the bus and scheduling of the individual processing
units. A fully pipelined architecture such as the example in Figure 13.13 has the potential to give
high performance due to pipelined execution by the separate functional units. However, this
type of architecture may require significant redesign in order to support a different coding
standard or a new optional coding mode.

A further consideration for a hardware design is the partitioning between the dedicated
hardware and the 'host' processor. A 'co-processor' architecture such as that described in the
DirectX VA framework (see Chapter 13) implies close interworking between the host and
the hardware on a macroblock-by-macroblock basis. An alternative approach is to move
more operations into hardware, for example by allowing the hardware to process a complete
frame of video independently of the host.

### 13.4.3   Designing the Functional Blocks

The choice of design for each functional block depends on the design goals (e.g. low area
and/or power consumption vs. high performance) and to a certain extent on the choice of
architecture. A 'common bus'-type architecture may lend itself to the reuse of certain
'expensive' processing elements. Basic operations such as multiplication may be reused by
several functional blocks (e.g. DCT and quantise). With the 'pipelined' type of architecture,
individual modules do not usually share processing elements and the aim is to implement
each function as efficiently as possible, for example using slower, more compact distributed
designs such as the distributed arithmetic architecture described in Chapter 7.

In general, regular, modular designs are preferable both for ease of design and efficient
implementation on the target platform. For example, a motion estimation algorithm that
maps to a regular hardware design (e.g. hierarchical search) may be preferable to less regular
algorithms such as nearest-neighbours search (see Chapter 6).

### 13.4.4   Testing

Testing and verification of a hardware CODEC can be a complicated process, particularly
since it may be difficult to test with 'real' video inputs until a hardware prototype is
available. It may be useful to develop a software model that matches the hardware design to

assist in generating test vectors and checking the results. A real-time test bench, where a hardware design is implemented on a reprogrammable FPGA in conjunction with a host system and video capture/display capabilities, can support testing with a range of real video sequences.

VLSI video CODEC design approaches and examples have been reviewed[9-12] and two specific design case studies presented.[11,12]

## 13.5 SUMMARY

The design of a video CODEC depends on the target platform, the transmission environment and the user requirements. However, there are some common goals and good design practices that may be useful for a range of designs. Interfacing to a video CODEC is an important issue, because of the need to efficiently handle a high bandwidth of video data in real time and because flexible control of the CODEC can make a significant difference to performance. There are many options for partitioning the design into functional blocks and the choice of partition will affect the performance and modularity of the system. A large number of alternative algorithms and designs exist for each of the main functions in a video CODEC. A good design approach is to use simple algorithms where possible and to replace these with more complex, optimised algorithms in performance-critical areas of the design. Comprehensive testing with a range of video material and operating parameters is essential to ensure that all modes of CODEC operation are working correctly.

## REFERENCES

1. I. Richardson, K. Kipperman and G. Smith, 'Video coding using digital signal processors', DSP World Fall Conference, Orlando, 1999.
2. J. McVeigh et al., 'A software-based real-time MPEG-2 video encoder', *IEEE Trans. CSVT*, **10**(7), October 2000.
3. S. Akramullah, I. Ahmad and M. Liou, 'Optimization of H.263 video encoding using a single processor computer', *IEEE Trans. CSVT*, **11**(8), August 2001.
4. B. Erol, F. Kossentini and H. Alnuweiri, 'Efficient coding and mapping algorithms for software-only real-time video coding at low bit rates', *IEEE Trans. CSVT*, **10**(6), September 2000.
5. N. Yung and K. Leung, 'Spatial and temporal data parallelization of the H.261 video coding algorithm', *IEEE Trans. CSVT*, **11**(1), January 2001.
6. K. Leung, N. Yung and P. Cheung, 'Parallelization methodology for video coding–an implementation on the TMS320C80', *IEEE Trans. CSVT*, **10**(8), December 2000.
7. A. Hamosfakidis, Y. Paker and J. Cosmas, 'A study of concurrency in MPEG-4 video encoder', *Proceedings of IEEE Multimedia Systems'98*, Austin, Texas, July 1998.
8. S. D. Kim, S. K. Jang, J. Lee, J. B. Ra, J. S. Kim, U. Joung, G. Y. Choi and J. D. Kim, 'Efficient hardware–software co-implementation of H.263 video CODEC', *Proc. IEEE Workshop on Multimedia Signal Processing*, pp. 305–310, Redondo Beach, Calif., 7–9 December 1998.
9. P. Pirsch, N. Demassieux and W. Gehrke, 'VLSI architectures for video compression–a survey', *Proceedings of the IEEE*, **83**(2), February 1995.
10. P. Pirsch and H. -J. Stolberg, 'VLSI implementations of image and video multimedia processing systems', *IEEE Transactions on Circuits and Systems for Video Technology*, **8**(7), November 1998, pp. 878–891.

11. P. Pirsch and H. -J. Stolberg, 'VLSI implementations of image and video multimedia processing systems', *IEEE Transactions on Circuits and Systems for Video Technology*, **8**(7), November 1998, pp. 878–891.
12. A. Y. Wu, K. J. R. Liu, A. Raghupathy and S. C. Liu, *System Architecture of a Massively Parallel Programmable Video Co-Processor*, Technical Report ISR TR 95-34, University of Maryland, 1995.

# 14

# Future Developments

## 14.1  INTRODUCTION

This book has concentrated on the design of video CODECs that are compatible with current standards (in particular, MPEG-2, MPEG-4 and H.263) and on the current 'state of the art' in video coding technology.[1] Video coding is a fast-moving subject and current research in the field moves beyond the bounds of the international standards; at the same time, improvements in processing technology will soon make it possible to implement techniques that were previously considered too complex. This final chapter reviews trends in video coding standards, research and platforms.

## 14.2  STANDARDS EVOLUTION

The ISO MPEG organisation is at present concentrating on two main areas: updates to existing standards and a new standard, MPEG-21. MPEG-4 is a large and complex standard with many functions and tools that go well beyond the basic H.263-like functionality of the popular 'simple profile' CODEC. It was originally designed with continual evolution in mind: as new techniques and applications become mature, extra tools and profiles continue to be added to the MPEG-4 set of standards. Recent work, for example, has included new profiles that support some of the emerging Internet-based applications for MPEG-4. Some of the more advanced elements of MPEG-4 (such as sprite coding and model-based coding) are not yet widely used in practice, partly for reasons of complexity. As these elements become more popular (perhaps due to increased processor capabilities), it may be that their description in the standard will need to be modified and updated.

MPEG-21[2] builds on the coding tools of MPEG-4 and the content description tools of the MPEG-7 standard to provide a 'framework' for multimedia communication. The MPEG committee has moved beyond the details of coding and description to an ambitious effort to standardise aspects of the complete multimedia 'delivery chain', from creation to 'consumption' (viewing or interacting with the data). This process may include the standardisation of new coding and compression tools.

The *Video Coding Experts Group* of the ITU continues to develop the H.26x series of standards. The recently added Annexes V, W and X of H.263 are expected to be the last major revisions to this standard. The main ongoing effort has to finalise the first version of H.26L: the core tools of the standard (described in Chapter 5) are reasonably well defined, but there is further work required to convert these into a published international standard. The technical aspects of H.26L were scheduled to be finalised during 2001. However, there is now an initiative between MPEG and VCEG to jointly develop a new coding standard based on H.26L$^3$.

## 14.3 VIDEO CODING RESEARCH

Video coding technology remains a very active area for researchers. Research in this field falls into two main categories, 'applied' research into the practical implementation of established video coding techniques and 'speculative' research into new and emerging coding algorithms. As a guide to the subjects that are currently popular in the research community, it is interesting to examine the papers presented at the 2001 Picture Coding Symposium (a specialist forum for image and video coding research). The total of 110 papers included:

- 22 papers on the implementation and optimisation of the popular block DCT-based video coding standards;
- 11 papers on transmission issues;
- 7 papers on quality measurement and quality metrics;
- 22 papers on content-based and object-based coding (including MPEG-4 object-based coding);
- 5 papers on wavelet-based coding of video sequence;
- 5 papers on coding of 3D/multi-view video.

(Note that some papers were difficult to categorise.) This cross-section of topics implies that much of the current research effort focuses on practical implementation issues for the popular block-based coding standards. The object-based functions of the MPEG-4 standard attract a lot of research interest and the feeling is that there are still a number of practical problems to solve (such as reliable, automatic segmentation of video scenes into video object planes) before these tools become widely adopted by the multimedia industry. A surprisingly small number of papers were presented on 'blue sky' research into novel coding methods. It is important to research and develop the next generation of video coding algorithms; at the same time, there is clearly a lot of scope for improving and optimising the current generation of coding technology.

## 14.4 PLATFORM TRENDS

Chapter 12 summarised the key features of a range of platforms for video CODEC implementation. There is some evidence of convergence between some of these platforms; for example, PC processor manufacturers continue to add instructions and features that were formerly encountered in special-purpose video or media processors. However, it is likely that there will continue to be distinct classes of platform for video coding, possibly along the following lines:

1. PC processors with media processing functions and increasing use of hardware co-processing (e.g. in video display cards).

2. More 'streamlined' processors (e.g. embedded processors with internal or external multimedia support, or media processors) for embedded multimedia applications.

3. Dedicated hardware CODECs (with limited programmability) for efficient implementation of 'mass-market' applications such as digital TV decoding.

There is still a place in the market for dedicated hardware designs but at the same time there is a trend towards flexible, embedded designs for new applications such as mobile multimedia. The increasing use of 'system on a chip' (SoC) techniques, with which a complex IC design can be rapidly put together from Intellectual Property building blocks, should make it possible to quickly reconfigure and redesign a 'dedicated' hardware CODEC. This will be necessary if dedicated designs are to continue to compete with the flexibility of embedded or general-purpose processors.

## 14.5 APPLICATION TRENDS

Predicting future directions for multimedia applications is notoriously difficult. Few of the 'interactive' applications that were proposed in the early 1990s, for example, have gained a significant market presence. The largest markets for video coding at present are probably digital television broadcasting and DVD-video (both utilising MPEG-2 coding). Internet video is gaining popularity, but is hampered by the limited Internet connections experienced by most users. There are some signs that MPEG-4 coding for video compression, storage and playback may experience a boom in popularity similar to MPEG Layer 3 Audio ('MP3' audio). However, much work needs to be done on the management and protection of intellectual property rights before this can take place.

Video conferencing via the Internet (typically using the H.323 protocol family) is becoming more widely used and may gain further acceptance with increases in processor and connection performance. It has yet to approach the popularity of communication via voice, e-mail and text messaging. There are two application areas that are currently of interest to developers and communications providers, at opposite ends of the bandwidth spectrum:

1. Very low power, very low bandwidth video for hand-held mobile devices (one of the hoped-for 'killer applications' for the costly third-generation mobile networks). The challenge here is to provide usable, low-cost video services that could match the popularity of mobile telephony.

2. High bandwidth, high quality video coding for applications such as:
   (a) 'Immersive' video conferencing, for example displaying conference participants on a video 'wall' as if they were sitting across a table from each other. The eventual goal is a video conference meeting that is almost indistinguishable from a face-to-face meeting.

(b) High definition television (HDTV, approximately twice the resolution of ITU-R 601 'standard' digital television). Coding methods (part of MPEG-2) have been standardised for several years but this technology has not yet taken hold in the marketplace.

(c) Digital cinema offers an alternative to the reels of projector film that are still used for distribution and display of cinema films. There is currently an effort by the MPEG committee (among others) to develop standard(s) to support cinema-quality coding of video and audio. MPEG's requirements document for digital cinema[4] specifies 'visually lossless' compression (i.e. no loss should be discernible by a human observer in a movie theatre) of frames containing up to 16 million pixels at frame rates of up to 150 Hz. In comparison, an ITU-R 601 frame contains around 0.5 million pixels. Coding and decoding at cinema fidelity are likely to be extremely demanding and will pose some difficult challenges for CODEC developers.

An interesting by-product of the 'mainstream' video coding applications and standards is the growing list of new and innovative applications for digital video. Some examples include the use of 'live' video in computer games; video 'chat' on a large scale with multiple participants; video surveillance in increasingly hostile environments (such as in an oil well or inside the body of a patient); 3-D video conferencing; video conferencing for groups with special requirements (for example deaf users); and many others.

Early experiences have taught designers of digital video applications that an application will only be successful if users find it to be a usable, useful improvement over existing technology. In many cases the design of the user interface is as important as, or more important than, the efficiency of a video coding algorithm. Usability is a vital but often overlooked requirement for any new video-based application.

## 14.6  VIDEO CODEC DESIGN

The aim of this book has been to introduce readers to the concepts, standards, design techniques and practical considerations behind the design of video coding and communication systems. A question that is often raised is whether the huge worldwide effort in video coding research and development will continue to be necessary, since transmission bandwidths may perhaps reach the point at which compression becomes unnecessary.

Video and multimedia applications have only begun to make a significant impact on businesses and consumers since the late 1990s. Despite continued improvements in resources such as processing power, storage and bandwidth, these resources continue to be stretched by increasing demands for high-quality, realistic multimedia communications with more functionality. There is still a large gap between the expectations of the user and the capabilities of present-day video applications and this gap shows no sign of diminishing. As digital video increases its share of the market, consumer demands for higher-quality, richer multimedia services will continue to increase. Bridging the gap (providing better quality and functionality within the limits of bandwidth and processing power) requires, among other things, continued improvements in video CODEC design.

In the past, market researchers have overestimated the rate of take-up of multimedia applications such as digital TV and video conferencing and it remains to be seen whether there is a real demand for some of the newer video services such as mobile video. Some interesting trends (for example, the continued popularity of MJPEG video CODECs because of their design simplicity and inherent error resilience) imply that the video communications market is likely to continue to be driven more by user needs than by impressive research developments. This in turn implies that only some of the recent developments in video coding (such as object-based coding, content-based tools, media processors and so on) will survive. However, video coding will remain a core element of the growing multimedia communications market. Platforms, algorithms and techniques for video coding will continue to change and evolve. It is hoped that this book will help to make the subject of video CODEC design accessible to a wider audience of designers, developers, integrators and users.

# REFERENCES

1. T. Ebrahimi and M. Kunt, 'Visual data compression for multimedia applications: an overview', *Proceedings of the IEEE*, **86**(6), June 1998.
2. ISO/IEC JTC1/SC29/WG11 N4318, 'MPEG-21 overview', Sydney, July 2001.
3. ITU-T Q6/SG16 VCEG-L45, 'H.26L Test Model Long-term number 6 (TML-6) draft 0', March 2001.
4. ISO/IEC JTC1/SC29/WG11 N4331, 'Digital cinema requirements', Sydney, July 2001.

# Bibliography

1. Bhaskaran, V. and K. Konstantinides, *Image and Video Compression Standards: Algorithms and Architectures*, Kluwer, 1997.
2. Ghanbari, M. *Video Coding: An Introduction to Standard Codecs*, IEE Press, 1999.
3. Girod, B., G. Greiner and H. Niemann (eds), *Principles of 3D Image Analysis and Synthesis*, Kluwer, 2000.
4. Haskell, B., A. Puri and A. Netravali, *Digital Video: An Introduction to MPEG-2*, Chapman & Hall, 1996.
5. Netravali, A. and B. Haskell, *Digital Pictures: Representation, Compression and Standards*, Plenum Press, 1995.
6. Parhi, K. K. and T. Nishitani (eds), *Digital Signal Processing for Multimedia Systems*, Marcel Dekker, 1999.
7. Pennebaker, W. B. and J. L. Mitchell, *JPEG: Still Image Data Compression Standard,* Van Nostrand Reinhold, 1993.
8. Pennebaker, W. B., J. L. Mitchell, C. Fogg and D. LeGall, *MPEG Digital Video Compression Standard*, Chapman & Hall, 1997.
9. Puri, A. and T. Chen (eds), *Multimedia Systems, Standards and Networks*, Marcel Dekker, 2000.
10. Rao, K. R. and J. J. Hwang, *Techniques and Standards for Image, Video and Audio Coding*, Prentice Hall, 1997.
11. Rao, K. R. and P. Yip, *Discrete Cosine Transform*, Academic Press, 1990.
12. Riley, M. J. and I. G. Richardson, *Digital Video Communications*, Artech House, February 1997.

# Glossary

| | |
|---|---|
| 4 : 2 : 0 (sampling) | sampling method: chrominance components have half the horizontal and vertical resolution of luminance component |
| 4 : 2 : 2 (sampling) | sampling method: chrominance components have half the horizontal resolution of luminance component |
| 4 : 4 : 4 (sampling) | sampling method: chrominance components have same resolution as luminance component |
| API | application programming interface |
| arithmetic coding | coding method to reduce redundancy |
| artefact | visual distortion in an image |
| BAB | binary alpha block, indicates the boundaries of a region (MPEG-4 Visual) |
| baseline (CODEC) | a codec implementing a basic set of features from a standard |
| block matching | motion estimation carried out on rectangular picture areas |
| blocking | square or rectangular distortion areas in an image |
| B-picture | coded picture predicted using bidirectional motion compensation |
| channel coding | error control coding |
| chrominance | colour difference component |
| CIF | common intermediate format, a colour image format |
| CODEC | *CO*der/*DEC*oder pair |
| colour space | method of representing colour images |
| DCT | discrete cosine transform |
| DFD | displaced frame difference (residual image after motion compensation) |
| DPCM | differential pulse code modulation |
| DSCQS | double stimulus continuous quality scale, a scale and method for subjective quality measurement |
| DVD | digital versatile disk |
| DWT | discrete wavelet transform |
| entropy coding | coding method to reduce redundancy |
| error concealment | post-processing of a decoded image to remove or reduce visible error effects |
| field | odd- or even-numbered lines from a video image |
| flowgraph | pictorial representation of a transform algorithm (or the algorithm itself) |
| full search | a motion estimation algorithm |
| GOB | group of blocks, a rectangular region of a coded picture |
| GOP | group of pictures, a set of coded video images |
| H.261 | standard for video coding |
| H.263 | standard for video coding |
| H.26L | 'Long-term' standard for video coding |
| HDTV | high definition television |
| Huffman coding | coding method to reduce redundancy |

| | |
|---|---|
| HVS | human visual system, the system by which humans percieve and interpret visual images |
| inter-frame (coding) | coding of video frames using temporal prediction or compensation |
| interlaced (video) | video data represented as a series of fields |
| intra-frame (coding) | coding of video frames without temporal prediction |
| ISO | International Standards Organisation |
| ITU | International Telecommunication Union |
| ITU-R 601 | a colour video image format |
| JPEG | Joint Photographic Experts Group, a committee of ISO; also an image coding standard |
| JPEG-2000 | an image coding standard |
| KLT | Karnuhen–Loeve transform |
| latency | delay through a communication system |
| loop filter | spatial filter placed within encoding or decoding feedback loop |
| MCU | multi-point control unit, controls a multi-party conference |
| media processor | processor with features specific to multimedia coding and processing |
| memory bandwidth | Data transfer rate to/from RAM |
| MJPEG | System of coding a video sequence using JPEG intra-frame compression |
| motion compensation | prediction of a video frame with modelling of motion |
| motion estimation | estimation of relative motion between two or more video frames |
| motion vector | vector indicating a displaced block or region to be used for motion compensation |
| MPEG | Motion Picture Experts Group, a committee of ISO |
| MPEG-1 | a video coding standard |
| MPEG-2 | a video coding standard |
| MPEG-4 | a video coding standard |
| objective quality | visual quality measured by algorithm(s) |
| OBMC | overlapped block motion compensation |
| profile | a set of functional capabilities (of a video CODEC) |
| progressive (video) | video data represented as a series of complete frames |
| pruning (transform) | reducing the number of calculated transform coefficients |
| PSNR | peak signal to noise ratio, an objective quality measure |
| QCIF | quarter common intermediate format |
| QoS | quality of service |
| quantise | reduce the precision of a scalar or vector quantity |
| rate control | control of bit rate of encoded video signal |
| rate-distortion | measure of CODEC performance (distortion at a range of coded bit rates) |
| RGB | red/green/blue colour space |
| ringing (artefacts) | 'ripple'-like artefacts around sharp edges in a decoded image |
| RTP | real-time protocol, a transport protocol for real-time data |
| RVLC | reversible variable length code |
| scalable coding | coding a signal into a number of layers |
| short header (MPEG-4) | a coding mode that is functionally identical to H.263 ('baseline') |
| SIMD | single instruction multiple data |
| slice | a region of a coded picture |
| statistical redundancy | redundancy due to the statistical distribution of data |
| subjective quality | visual quality as perceived by human observer(s) |
| subjective redundancy | redundancy due to components of the data that are subjectively insignificant |
| sub-pixel (motion compensation) | motion-compensated prediction from a reference area that may be formed by interpolating between integer-valued pixel positions |

| | |
|---|---|
| test model | a software model and document that describe a reference implementation of a video coding standard |
| TSS | three-step search, a motion estimation algorithm |
| VCA | variable complexity algorithm |
| VCEG | Video Coding Experts Group, a committee of ITU |
| video packet (MPEG-4) | coded unit suitable for packetisation |
| video processor | processor with features specific to video coding and processing |
| VLC | variable length code |
| VLD | variable length decoder |
| VLE | variable length encoder |
| VLIW | very long instruction word |
| VLSI | very large scale integrated circuit |
| VO (MPEG-4) | video object |
| VOP (MPEG-4) | video object plane |
| VQEG | Video Quality Experts Group |
| YCrCb | luminance/red chrominance/blue chrominance colour space |

# Index